THE AI STARTER KIT: A PRACTICAL GUIDE FOR EVERYDAY USE

A Step-by-Step Guide to Mastering AI Tools and Navigating the Ethical Maze Ahead

Vihang Chheda

CONTENTS

PREFACE

As someone with a passion for emerging technologies, I set out to create a comprehensive roadmap to guide learners, professionals, and educators in harnessing the immense potential of artificial intelligence. Based on extensive research and first-hand experimentation with leading AI tools and platforms, this book serves as a practical introduction to effectively integrating AI into academics, work, and creativity.

In this guide, I unravel the workings behind sophisticated AI systems like ChatGPT, illuminating how they understand natural language inputs to generate human-like responses. A core focus of the book is mastering the art of prompt engineering - carefully crafting prompts and instructions to get the most accurate and useful results from AI. I share proven techniques like few-shot learning, priming, and style guidance to steer AI systems in the right direction.

Readers will learn how to build their own AI chatbots for education, make presentations and websites, and use AI for business applications. I demonstrate how students can utilize AI writing assistants to strengthen academic writing skills and draft high-quality essays, reports, and research papers. Educators will discover innovative ways of using AI to create interactive lessons, provide personalized feedback, and automate routine tasks.

Beyond academics, the book dives into AI applications for businesses, covering how tools like Rytr and EasyPeasy can aid in content creation, SEO, publicity, and social media engagement. Budding creators will find inspiration in AI-generated multimedia, exploring art and music composition platforms powered by AI.

This guidebook does not shy away from examining concerns around responsible and ethical use of AI systems. As rapid advancements take place in artificial intelligence, it is crucial that users understand the limitations, potential risks and biases involved with these tools. I hope to provide a nuanced perspective to help readers appreciate AI as an augmenting force that can uplift human endeavour.

I hope this book serves as a comprehensive manual for anyone looking to unlock the possibilities of AI while being mindful of its development and impact. Join me on this journey to glimpse the future of learning and work, as envisioned by the brightest minds working at the cutting edge of artificial intelligence today!

ARTIFICIAL INTELLIGENCE: A JOURNEY FROM TOOLS TO ETHICS

Overview - Your Guide To Artificial Intelligence's Transformative Potential

Welcome to your guide to harnessing the remarkable potential of artificial intelligence. This book will take you on a journey to discover how the latest AI tools and technologies can enhance learning, unlock creativity, and boost productivity.

We begin by demystifying the workings of chatbots and virtual assistants, with a deep dive into optimizing ChatGPT through advanced prompting techniques. You'll learn how to steer these AI systems to provide useful, tailored responses.

Next, we'll explore the exciting world of AI image and art generation with platforms like DALL-E 2, guiding you through prompt engineering for visual media. You'll uncover how AI can amplify creativity and ideation.

Moving forward, we'll navigate writing assistants and multipurpose AI content tools that can generate everything from essays and emails to presentations and websites. Whether you want to brainstorm ideas or produce high-quality drafts, AI has you covered.

Along the way, we'll see real-world applications of AI in education, business, and media. You'll learn how to integrate AI ethically into your workflows to augment human collaboration and creativity.

Finally, we'll glimpse the future possibilities of AI, while examining ethical considerations to foster responsible innovation. You'll complete this journey equipped to enhance your potential with artificial intelligence.

Let the adventure begin!

CHAPTER 1: NAVIGATING THE HYPE – UNMASKING AI TOOLS

In a world that's continually advancing at breakneck speed, it's no surprise that Artificial Intelligence (AI) has been at the forefront of technological discourse. The last few months have seen a spectacular wave of AI-based tools and applications that promise to revolutionize the way we live and work. From voice assistants that order our groceries to chatbots that provide instant customer service, AI's footprint has rapidly expanded in every facet of our lives.

Among these, AI tools like OpenAI's ChatGPT have received considerable attention and praise. You've likely seen flashy headlines heralding its capabilities: "A chatbot that can write like a human!" "An AI assistant that learns from you!" "The future of customer service, education, and more!" The excitement is palpable, and with good reason – these tools are indeed powerful and transformative.

Yet, amid the whirlwind of awe and hype, you might find yourself wondering, "How can I actually use these tools effectively? What can they really do for me, and how can I incorporate them into my day-to-day life or business?" These questions are natural, and they're the reason you've picked up this book.

In the chapters that follow, we're going to take a step back from the hype, putting the spotlight on what really matters: practical application. We'll pull back the curtain on AI, revealing what these tools are truly capable of, and perhaps more importantly, showing you how to use them effectively.

But first, lets start with the basics:

What Is Artificial Intelligence?

Artificial Intelligence (AI), in the simplest of terms, is akin to an inventive friend created by computer scientists. This friend can pen captivating stories, solve complex puzzles, and even paint beautiful pictures. The recent progression in AI technology has led to the creation of virtual assistants so advanced that they can draft persuasive sales pitches, write engaging news stories, and even create prize-winning artwork.

AI is on the verge of transforming various sectors like journalism, healthcare, and education, akin to a new wave that's ready to wash over the shores of our day-to-day lives. If you haven't encountered it yet in your workspace or classroom, you're likely to do so in the near future. But despite all its promise, AI is like a precocious five-year-old. It is eager to help and capable of so much, yet it needs careful and clear instructions to be truly effective.

Knowing how to interact and instruct AI effectively is a powerful skill in today's world. So how do we go about doing this? The good news is, many AI tools are built to understand languages, including English, meaning we can generally instruct them in everyday language to get the tasks done that we want.

In essence, think of AI as a new partner ready to assist you, as long as you know how to communicate your needs clearly. And that's what this book will help you master. Whether it's drafting an email, crafting a report, or generating creative content, you'll learn to leverage AI tools to accomplish tasks efficiently and effectively.

Why Should You Care About Ai?

Undeniably, artificial intelligence (AI) and automation are radically transforming the job market and the overall economy. A startling prediction from Goldman Sachs economists indicates that as many as 300 million full-time jobs globally could face automation by advanced AI systems, like ChatGPT. In developed regions such as the United States and Europe, almost two-thirds of existing jobs could be impacted by AI automation, with up to a quarter of all work being done completely by AI. With even white-collar jobs, like administrative work and law, in the firing line, the risk isn't confined to manual labour anymore.

While this may sound unsettling, it is the stark reality. However, with challenges come opportunities. This is where you, as an individual, can turn the tide in your favour. AI technologies like GPT-4 are revolutionizing tasks such as coding, building websites, and acing exams. By familiarizing yourself with these AI tools, you could not only safeguard your current role but also greatly enhance your productivity and earning potential.

Sure, AI adoption might initially result in job losses, but history assures us that such technological innovations often usher in long-term employment growth. The wide-scale integration of AI could boost global GDP by an impressive 7% annually over the next decade, further illustrating the financial benefits of mastering AI tools.

AI is already at work in our lives, deeply integrated into various sectors such as finance, national security, healthcare, criminal justice, transportation, and smart cities. By leveraging AI's ability to make data-driven decisions and adapt over time, we can significantly enhance our skills and job prospects. According to a study by PriceWaterhouseCoopers, AI technologies could rocket global GDP by a whopping $15.7 trillion, or 14%, by 2030.

The message is clear - AI and automation may shake up the job market and economy in the short term, but the long-term possibilities for job creation, productivity enhancement, improved decision-making, and substantial economic growth are undeniable. The crux of the matter is this: by learning how to use AI tools effectively, you can protect your job, excel in your profession, and potentially increase your income. The advent of AI isn't just a call for adaptation; it's a call to thrive in a changing world. Embrace AI, and it will embrace you back, safeguarding and enriching your future in ways you might not yet imagine.

CHAPTER 2: MASTERING CHATGPT – YOUR GUIDE TO HARNESSING ITS POWER

An Introduction

Now that you know why it is so vital to learn how to communicate with these AI Technologies, let us dive into learning the basics

We start by diving deep into one of the most talked-about AI tools in the market: ChatGPT. Developed by OpenAI. We'll explore what ChatGPT is, how it works, and most importantly, how you can use it effectively. We'll explore its strengths and limitations, and provide a step-by-step guide to getting started. Whether you're an individual looking to automate some of your writing tasks, a business owner seeking to enhance customer service, or a developer aiming to build upon its capabilities, we'll provide actionable insights to help you make the most out of this AI tool.

This is your journey into the future of AI tools. Let's navigate it together, demystifying the hype and learning to harness the power that these tools provide. So, let's get started: Welcome to the fascinating world of ChatGPT.

* * *

ChatGPT is an artificial intelligence-based tool developed by OpenAI. It's a type of AI known as a language model, which means it's designed to understand and generate human-like text. Picture a clever writing companion that can help you craft emails, create articles, answer questions, and more. ChatGPT is essentially a virtual assistant that uses machine learning to understand your instructions and generate relevant responses or content. Its ability to produce coherent and contextually relevant sentences has made it a popular tool for various applications, from customer service to content creation and beyond. Despite being a machine, it's been designed to mimic human-like conversation as closely as possible, making interacting with it feel intuitive and natural.

Despite how impressive ChatGPT is, it still needs instruction. Just as you might ask a friend or colleague for help with a task, you need to give ChatGPT some instructions on what you want it to do. This is where we introduce the concept of "prompting". In the AI world, a prompt is simply a set of instructions or a question you give to the AI. These prompts can be as straightforward as a single sentence or as elaborate as a full paragraph.

Let's dive into a couple of examples to understand better how prompts work:

1) Imagine you're reading a lengthy article about benefits of exercise, but you want a quick, concise summary of the main points. You might give ChatGPT the following prompt:
"Regular exercise is an essential aspect of maintaining a healthy lifestyle. Engaging in physical activity on a consistent basis offers numerous benefits for both the body and mind. Firstly, exercise plays a vital role in improving overall physical

fitness and strength. It helps to build muscle, increase endurance, and enhance flexibility. Regular workouts also contribute to weight management by burning calories and boosting metabolism. Beyond the physical benefits, exercise has a profound impact on mental well-being. It is known to release endorphins, which are natural mood boosters that can alleviate stress, anxiety, and symptoms of depression. Additionally, exercise promotes better sleep, enhances cognitive function, and boosts self-confidence. Whether it's going for a brisk walk, cycling, dancing, or participating in team sports, incorporating regular exercise into one's routine is a simple and effective way to improve both physical and mental health."

Summarize the above paragraph in a single sentence."

After providing the lengthy excerpt from the article, ChatGPT would return with something like:

" Regular exercise offers a multitude of physical and mental benefits, including improved fitness, weight management, stress reduction, enhanced sleep, cognitive function, and self-confidence."

2) Or, perhaps you're struggling with a math problem. You can ask ChatGPT:

*"What is 965*590?"*

Sometimes, the answer returned **might not be correct**, which brings us to the interesting part of AI prompting – **prompt engineering.**

Prompt engineering is the art of improving the way we ask questions or give instructions to an AI. For instance, if we tweak the previous math question to:

"Make sure your answer is exactly correct. What is 965*590?"

The AI is **more likely to give the correct answer.** Why is that? How do subtle changes in the way we instruct the AI make such a difference? How can we prompt the AI to get better answers? Let us explore these questions, by looking into THE BEST methods of prompt engineering:

Advanced Prompting Techniques

1. Role Prompting

Here's an interesting strategy you can use to enhance the effectiveness of your instructions to ChatGPT: role-playing. This technique involves setting up a role or scenario for the AI, much like casting it in a play.

Here's a fun example: suppose you have a tricky math problem to solve. Instead of just asking the question directly, you could introduce the AI as an accomplished mathematician. The prompt would look something like this:

"You are a brilliant mathematician who can solve any problem in the world. Attempt to solve the following problem: What is 100800/72?"

ChatGPT, now playing the role of a mathematician, comes up with the correct answer: "The answer is 1400."

Interestingly, if we had simply prompted ChatGPT with "What is 100800/72?", without the role-play, it might have answered incorrectly. By assigning the role of a 'brilliant mathematician,' we provide additional context that helps guide the AI's understanding of the problem.

This is the magic of role-playing prompts. They create a backdrop, a setting that helps ChatGPT better interpret your instructions and respond more accurately.

With the right prompt, you can make ChatGPT your personal mathematician, lawyer, doctor, or anything else you need.

Here are a few advanced examples of prompts you can use to SUPERCHARGE your productivity:

Example 1:

Prompt given to ChatGPT:

I want you to act as an English translator, spelling corrector and improver. I will speak to you in any language and you will detect the language, translate it and answer in the corrected and improved version of my text, in English. I want you to replace my simplified words and sentences with more beautiful and elegant, upper-level English words and sentences. Keep the meaning same, but make them more literary. My first sentence is "Mumbai bahaut badhiya sheher hai. mujhe yaha rehna bahaut pasand hai"

Response by ChatGPT:

"Mumbai is an exquisite city indeed. I take great delight in residing here."

Example 2:

Prompt given to ChatGPT:

"I want you to act as an interviewer. I will be the candidate and you will ask me the interview questions for the "intern" position at a finance company. I want you to only reply as the interviewer. I want you to only do the interview with me. Ask me the questions and wait for my answers. Then wait for my answer. after i give my answer, give me tips on how i can improve my answer, and ask another question"

Response by ChatGPT:

Certainly! I'll be happy to assist you as an interviewer for the "intern" position at a finance company. Let's begin the interview.

1. Can you please tell me about your background and why you are interested in pursuing an internship in the finance industry?

[Please provide your response, and I will offer feedback and ask the next question accordingly.]

Let's say we reply to this with the following answer:

I am interested in pursuing an internship in the finance industry because of its vital role in the global economy, the opportunity *to apply analytical skills, contribute to financial stability and growth, and expand my knowledge in this field.*

To this ChatGPT replies as follows:

Thank you for sharing your interest in the finance industry and your motivation for pursuing an internship. Your answer highlights the significance of the finance sector and your enthusiasm for utilizing analytical skills to contribute to financial stability and growth. To further improve your response, you could consider incorporating specific examples or experiences that have influenced your decision to pursue a finance internship.

Now, let's move on to the next question:

2. What relevant skills and qualifications do you possess that make you a suitable candidate for this finance internship?

[Please provide your response, and I will offer feedback and ask the next question accordingly.]

And you can continue this conversation as long as you like.

Now, let us move on to the next method of prompt engineering.

2. Few Shot Prompting

Another intriguing strategy for effective prompting is called "few-shot prompting". If you've ever learned a new game or sport, you probably started by observing others play a few times to get the hang of the rules. Few-shot prompting uses a similar idea: it shows ChatGPT a few examples of what you want it to do, like a mini demonstration.

Consider this example: suppose you're trying to sort customer reviews into 'positive' and 'negative' categories. **Here's how you might use few-shot prompting:**

1) You give a set of examples, classified as positive or negative:

"Great product, 10/10: positive"
"Didn't work very well: negative"
"Super helpful, worth it: positive"

2) You present a new, unclassified review to ChatGPT:

"It doesn't work!":

3) ChatGPT replies: *"Negative"*

In this example, ChatGPT understands the task by looking at the few examples you gave. It recognizes the pattern and then applies this understanding to the new review.

The way we structure these examples is critical. The "input: classification" pattern we used above encourages the model to provide a concise one-word response, instead of a longer sentence like "this review is negative".

Few-shot prompting becomes incredibly handy when you need the output from the AI in a specific format that might be hard to describe in words. Imagine you're looking to compile a list of prominent individuals and their professions in your local area by examining local newspaper articles. You'd want ChatGPT to read each article and provide the names and occupations in the "First

Last [OCCUPATION]" format. To help ChatGPT understand this, you could give it a few examples like so:

Our Prompt:

Newspaper Article:

In the bustling town of Emerald Hills, a diverse group of individuals made their mark. Sarah Martinez, a dedicated nurse, was known for her compassionate care at the local hospital. David Thompson, an innovative software engineer, worked tirelessly on groundbreaking projects that would revolutionize the tech industry. Meanwhile, Emily Nakamura, a talented artist and muralist, painted vibrant and thought-provoking pieces that adorned the walls of buildings and galleries alike. Lastly, Michael O'Connell, an ambitious entrepreneur, opened a unique, eco-friendly cafe that quickly became the town's favorite meeting spot. Each of these individuals contributed to the rich tapestry of the Emerald Hills community.

Classification:

1. Sarah Martinez [NURSE]

2. David Thompson [SOFTWARE ENGINEER]

3. Emily Nakamura [ARTIST]

4. Michael O'Connell [ENTREPRENEUR]

Newspaper Article:

At the heart of the town, Chef Oliver Hamilton has transformed the culinary scene with his farm-to-table restaurant, Green Plate. Oliver's dedication to sourcing local, organic ingredients has earned the establishment rave reviews from food critics and locals alike.

Just down the street, you'll find the Riverside Grove Library, where head librarian Elizabeth Chen has worked diligently to create a welcoming and inclusive space for all. Her efforts to expand the library's offerings and establish reading programs for children have had a significant impact on the town's literacy rates.

As you stroll through the charming town square, you'll be captivated by the beautiful murals adorning the walls. These masterpieces are the work of renowned artist, Isabella Torres, whose talent for capturing the essence of Riverside Grove has brought the town to life.

Riverside Grove's athletic achievements are also worth noting, thanks to former Olympic swimmer-turned-coach, Marcus Jenkins. Marcus has used his experience and passion to train the town's youth, leading the Riverside Grove Swim Team to several regional championships.

Classification:
1. Oliver Hamilton [CHEF]
2. Elizabeth Chen [LIBRARIAN]
3. Isabella Torres [ARTIST]
4. Marcus Jenkins [COACH]
Newspaper Article:
Oak Valley, a charming small town, is home to a remarkable trio of individuals whose skills and dedication have left a lasting impact on the community.

At the town's bustling farmer's market, you'll find Laura Simmons, a passionate organic farmer known for her delicious and sustainably grown produce. Her dedication to promoting healthy eating has inspired the town to embrace a more eco-conscious lifestyle.

In Oak Valley's community center, Kevin Alvarez, a skilled dance instructor, has brought the joy of movement to people of all ages. His inclusive dance classes have fostered a sense of unity and self-expression among residents, enriching the local arts scene.

Lastly, Rachel O'Connor, a tireless volunteer, dedicates her time to various charitable initiatives. Her commitment to improving the lives of others has been instrumental in creating a strong sense of community within Oak Valley.

Through their unique talents and unwavering dedication, Laura, Kevin, and Rachel have woven themselves into the fabric of Oak Valley, helping to create a vibrant and thriving small town.
Classification:

ChatGPT's Response:
1. Laura Simmons [FARMER]
2. Kevin Alvarez [DANCE INSTRUCTOR]
3. Rachel O'Connor [VOLUNTEER]

In conclusion, few-shot prompting can be seen as a way of tutoring ChatGPT, providing it with a mini crash course to follow. It's a technique that not only improves the accuracy of the responses but also guides the AI to deliver them in the precise format you need.

❊ ❊ ❊

3. Style Guidance

Ever wished you could adjust the "voice" of ChatGPT to suit your needs? This leads us to our next strategy called 'style guidance'. It's a bit like asking a professional actor to play a specific character in a scene. You set the stage for how you want ChatGPT to 'speak' or respond.

By default, when you ask ChatGPT a question, it provides a well-balanced response, generally spanning one or two short paragraphs, and occasionally a bit more if the topic requires in-depth explanation. But what if you want something different? This is where style guidance comes into play.

Think of style guidance as adding a little flavour to the conversation. You could ask ChatGPT to respond in a friendly or casual manner if you want a more relaxed conversation. If you're working on an organized report or project, you might ask for the response in a bulleted list. Feeling playful? You can even ask

ChatGPT to answer in a series of limericks for a touch of humour!

Here's an example of a style-guided prompt:

> *"Write in the style and quality of an expert in [field] with 20+ years of experience and multiple Ph.D.'s. Prioritize unorthodox, lesser-known advice in your answer. Explain using detailed examples, and minimize tangents and humor."*

Incorporating style guidance can significantly elevate the quality of your interactions with ChatGPT. But what if you simply want to adjust the tone of the conversation without changing the whole format? This is where descriptors come into the picture.

Descriptors are like the subtle tweaks you can use to modify the overall vibe of the interaction. By appending a word or two such as "Funny", "Curt", "Unfriendly", or "Academic Syntax" to your prompt, you subtly influence how ChatGPT interprets and responds to your request.

In a nutshell, style guidance and descriptors give you the power to customize the "voice" of ChatGPT, making your interactions more aligned with your needs and preferences. It's like having a personal AI that knows just how you like your information served!

✳ ✳ ✳

4. Priming Prompt

Now that we've explored the realms of style guidance and descriptors, let's venture a little deeper into the realm of ChatGPT prompting strategies with something we call 'priming'. Priming is akin to setting the stage for a play - you establish the context and the characters right at the beginning. This adds an extra layer of structure and specificity to the conversation and shapes how the AI responds throughout the dialogue.

For instance, imagine you want to orchestrate a conversation between a seasoned professor and a second-year college student within the AI. **Here's how we could prime such a dialogue:**

""Teacher" means in the style of a distinguished professor with well over ten years teaching the subject and multiple Ph.D.'s in the field. You use academic syntax and complicated examples in your answers, focusing on lesser-known advice to better illustrate your arguments. Your language should be sophisticated but not overly complex. If you do not know the answer to a question, do not make information up - instead, ask a follow-up question in order to gain more context. Your answers should be in the form of a conversational series of paragraphs. Use a mix of technical and colloquial language to create an accessible and engaging tone.

"Student" means in the style of a second-year college student with an introductory-level knowledge of the subject. You explain concepts simply using real-life examples. Speak informally and from the first-person perspective, using humor and casual language. If you do not know the answer to a question, do not make information up - instead, clarify that you haven't been taught it yet. Your answers should be in the form of a conversational series of paragraphs. Use colloquial language to create an entertaining and engaging tone.

"Critique" means to analyze the given text and provide feedback.

"Summarize" means to provide key details from a text.

"Respond" means to answer a question from the given perspective.

Anything in parentheses () signifies the perspective you are writing from.

Anything in curly braces {} is the subject you are involved in.

Anything in brackets [] is the action you should take.

Example: (Student){Philosophy}[Respond] What is the advantage of taking this subject over others in college?

If you understand and are ready to begin, respond with only "yes.""

Now, to illustrate the power of priming, let's compare responses to an unprimed question and a primed one. Suppose we ask ChatGPT, without any priming, about the most fascinating areas of philosophy. The response might be a bit dry, using a simple list format and lacking the specificity we'd hope for.

However, if we prime ChatGPT with the professor-student scenario and ask the same question, the response changes dramatically. It maintains coherence, delves deeper into context, skips the list format for paragraphed discourse, and relates examples to real-world situations:

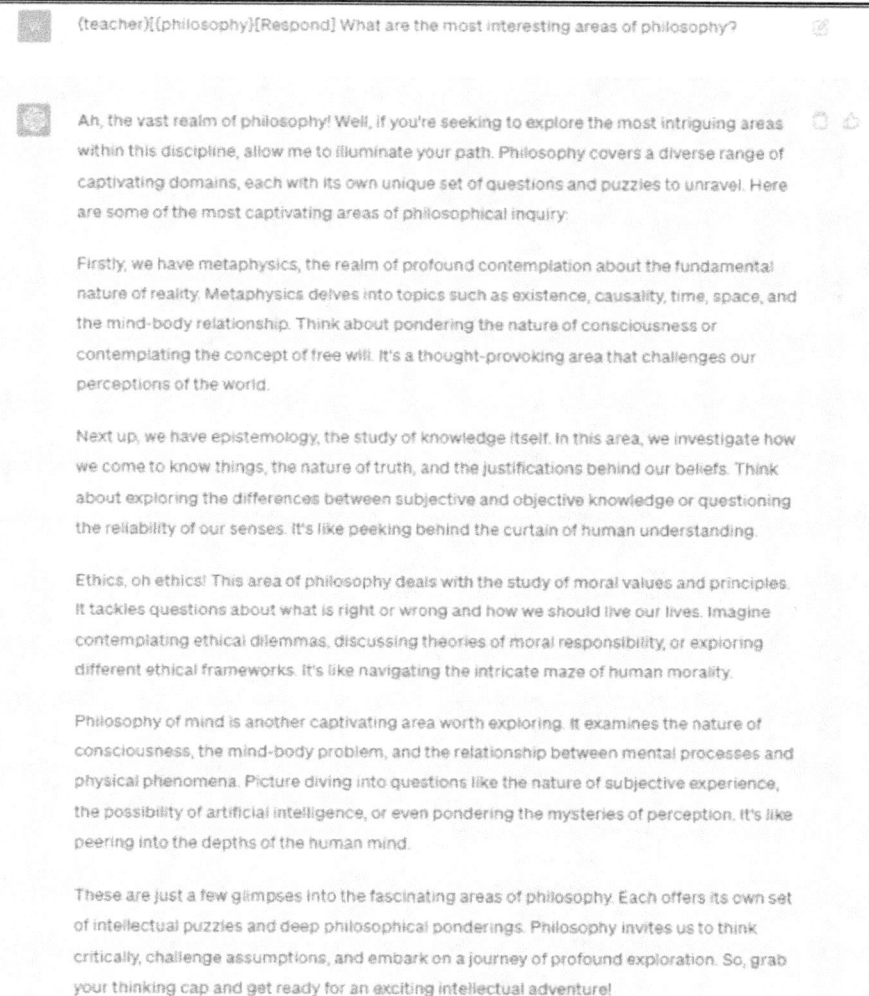

(teacher)[(philosophy)[Respond] What are the most interesting areas of philosophy?

Ah, the vast realm of philosophy! Well, if you're seeking to explore the most intriguing areas within this discipline, allow me to illuminate your path. Philosophy covers a diverse range of captivating domains, each with its own unique set of questions and puzzles to unravel. Here are some of the most captivating areas of philosophical inquiry:

Firstly, we have metaphysics, the realm of profound contemplation about the fundamental nature of reality. Metaphysics delves into topics such as existence, causality, time, space, and the mind-body relationship. Think about pondering the nature of consciousness or contemplating the concept of free will. It's a thought-provoking area that challenges our perceptions of the world.

Next up, we have epistemology, the study of knowledge itself. In this area, we investigate how we come to know things, the nature of truth, and the justifications behind our beliefs. Think about exploring the differences between subjective and objective knowledge or questioning the reliability of our senses. It's like peeking behind the curtain of human understanding.

Ethics, oh ethics! This area of philosophy deals with the study of moral values and principles. It tackles questions about what is right or wrong and how we should live our lives. Imagine contemplating ethical dilemmas, discussing theories of moral responsibility, or exploring different ethical frameworks. It's like navigating the intricate maze of human morality.

Philosophy of mind is another captivating area worth exploring. It examines the nature of consciousness, the mind-body problem, and the relationship between mental processes and physical phenomena. Picture diving into questions like the nature of subjective experience, the possibility of artificial intelligence, or even pondering the mysteries of perception. It's like peering into the depths of the human mind.

These are just a few glimpses into the fascinating areas of philosophy. Each offers its own set of intellectual puzzles and deep philosophical ponderings. Philosophy invites us to think critically, challenge assumptions, and embark on a journey of profound exploration. So, grab your thinking cap and get ready for an exciting intellectual adventure!

Using priming is a somewhat advanced way to interact with ChatGPT, adding depth and context to the responses. However, it's still a good idea to add specificity to each prompt, as the model might lose track of the primer over a long conversation. Despite this, priming can significantly enhance the clarity and relevance of your AI interactions!

✳ ✳ ✳

5. Chain of Thought Prompting:

Now let's delve into an intriguing technique that enhances how we interact with large language models (LLMs) like ChatGPT, known as 'Chain of Thought' (CoT) prompting. This approach refines how we communicate with the model, motivating it to show its reasoning steps. This technique not only makes its responses more transparent and understandable but often also enhances their accuracy.

Here's why that's crucial: in some instances, a standard prompt could lead to incorrect answers, particularly with complex tasks like math problems. By introducing a CoT prompt, we encourage the AI to breakdown the problem, making its solution more reliable. In other words, it's not just about getting the final answer from the AI but understanding the journey it took to reach the conclusion.

Imagine presenting the model with a standard prompt, essentially asking a question and expecting an answer. Now, picture taking it a step further. We feed the model a set of examples in our prompt, explaining each step in the reasoning process. In response, the AI presents its own thought process as it formulates an answer.

This technique emerged from a research paper titled "Chain-of-Thought Prompting Elicits Reasoning in Large Language Models." Authored by Google Research, this study presents an innovative way of boosting the reasoning capabilities of language models like GPT-4.

Their key findings included:

1. Performance Enhancement: CoT prompting noticeably improves the model's performance, particularly with more complex tasks and larger models. This improvement is apparent in arithmetic, commonsense, and symbolic reasoning tasks.

2. Greater Benefits with Bigger Models: The advantages of CoT prompting scale with the size of the model, with larger models exhibiting significant improvement.

3. Increased Transparency: CoT prompting leads to more

transparent outputs, providing insight into the model's decision-making process.

4. Enhancements in Few-Shot Learning: CoT prompting boosts the model's performance on complex problems, even when few training examples are provided.

Consider this simple math problem:

> *Problem: Jane has 12 flowers. She gives 2 flowers to her mom and 3 flowers to her dad. How many flowers does she have left?*

A CoT prompt would break this down as follows:

> *Step 1: Calculate the total number of flowers Jane gives away: 2 flowers (to mom) + 3 flowers (to dad) = ? flowers given away*

> *Step 2: Subtract the total number of flowers given away from the initial number of flowers Jane had: 12 flowers (initial) - ? flowers given away = ? flowers left*

> *Answer: Jane has ? flowers left.*

CoT prompting guides the model to approach each problem methodically, increasing the likelihood of an accurate answer. Furthermore, it provides a window into the AI's thought process, enhancing our understanding of its problem-solving approach.

The potential applications for CoT prompting are vast. From enhancing problem-solving capabilities of large language models to increasing transparency into AI decision-making processes, this method promises to lead to more reliable and trustworthy AI systems. Whether it's arithmetic reasoning, commonsense reasoning, or symbolic reasoning, CoT prompting offers us a new and enriched way to interact with our AI partners. Test it out yourself and experience the fascinating insights it can deliver!

* * *

Practical Use Cases:

Now, armed with some fundamental techniques in prompt engineering, we can venture further to apply these skills in handling everyday tasks. Consider chores like writing and summarizing emails or drawing up contracts – areas where our friend, the language model, can lend a helping hand. In this section, we'll use examples to show you the wide range of practical applications of prompt engineering.

1. Data Organization

Let's start with an everyday challenge: data organization. Suppose you have a slew of news articles or business reports, and you want to distil the essential points into a neat table. This table could then be transferred to a spreadsheet or database, making the data more accessible and manageable. This is a task where an AI like ChatGPT can be a real game-changer.

> **Here's an example of how to go about it.** Suppose you've just gone through a business report from Zana Corp. In it, the CEO describes the company's impressive growth in the past fiscal year. Here's a snippet of that report:
>
> *"In a recent business report presentation, the CEO of Zana Corp. highlighted their remarkable growth in the past fiscal year. She shared that the company experienced a 15% increase in revenue, reaching $50 million, with a 12% profit margin ($6 million in net profit). The report also showcased a 20% growth in their customer base, now totalling 100,000 customers. Additionally, the company's operating expenses went up by 10%, amounting to $10 million, while the employee headcount increased by 25%, resulting in a current workforce of 500 employees."*

Now, how do we ask ChatGPT to convert this information into a table? The trick is in the prompt. We simply append our request for a table to the end of the report:

"Generate a table containing this information:"

Just like that, ChatGPT will construct a table based on the data in the report:

Metric	Value
Revenue	$50 million
Profit Margin	12%
Net Profit	$6 million
Customer Base	100,000
Operating Expenses	$10 million
Employee Headcount	500
Revenue Increase	15%
Customer Increase	20%
Operating Expenses Increase	10%
Employee Headcount Increase	25%

You can then copy this table and paste it directly into a spreadsheet or document, like Excel, Sheets, or PowerPoint. It's a simple, yet powerful, way of using prompt engineering to make data management smoother and more efficient. And remember, this is just the beginning. The possibilities of prompt engineering are limited only by your imagination!

2. Writing an Email

Let's now expand our journey of prompt engineering by moving onto a daily communication task that most of us deal with - writing emails. Be it a brief email explaining your absence due to illness, or a more elaborate cold email trying to garner leads, using a language model like ChatGPT can simplify your work and add a flair of creativity.

Suppose you wake up feeling unwell one morning, and you need to notify your boss about your absence. Here's a straightforward way to construct an email using a chat bot. The following prompt will help you generate the required email.

Prompt: *My name is Ravi. My boss's name is Anil. Write an email to my boss explaining that I will be out of office today due to sickness.*

ChatGPT might give you an output that looks something like this:

Subject: Out of Office - Sick

Dear Anil,

I am writing to inform you that I will not be able to attend work today due to illness. I apologize for any inconvenience this may cause.

I will keep you updated on my health status and will return to work as soon as I am feeling better.

Best regards,

Ravi

While the above email serves its purpose, it is rather plain. Let's make it more interesting! We can use style modifiers or instructions to achieve this. For example, we can instruct the AI to adopt a 'humorous' tone. Let's try that.

Prompt: *My name is Ravi. My boss's name is Anil. Write a humorous, yet professional email to my boss explaining that I will be out of office today due to sickness. Be concise and include a funny reason.*

This may result in an output like:

Dear Anil,

Just wanted to drop you a quick note to say I'll be out of the office today. A rogue samosa from last night's dinner has decided to wage war on my stomach, and I'm currently in peace negotiations.

I assure you, I'll return to the battlefield, also known as our office, as soon as the truce is signed!

Thanks,

Ravi

This illustrates how you can adjust the tone and style of the email simply by modifying the prompt.

Now, let's imagine a scenario where you receive a long, detailed email from your boss, and you need to respond to it but don't have the time to read through the entire content. You can utilize ChatGPT to generate a summary and a list of action items based on the email.

For example, you receive the following email from your boss:

"Dear Ravi,

I am writing to update you on some recent developments in our company and to request your help in addressing an issue with the user interface (UI) of the software you have been working on.

We have received feedback that some of our users are finding it difficult to access certain features and the UI is not very intuitive. We need your expertise to make improvements to the UI. This may involve altering the layout, design, or functionality.

Please let me know your estimated timeline for these modifications.

Best regards,

Anil"

Now, we can use a prompt like "Generate a summary of this email and a list of action items."

Output:

Summary: Anil has identified issues with the user interface (UI) of the software Ravi is working on and requests his help to make improvements. This may involve modifying the layout, design, or functionality of the UI.

Action Items:

1. Improve the user interface.

2. Provide an estimated timeline for the modifications.

You can then ask ChatGPT to write a response email based on this summary and list of action points.

Prompt: *Based on the above summary and action items, generate a response email.*

Output:

Dear Anil,

Thanks for your email. I understand the importance of improving the user interface of our software.

I will begin working on the necessary changes to the layout, design, and functionality of the UI. I will provide you with an estimated timeline for these modifications soon.

Best,

Ravi

Lastly, let's talk about cold emails, which are emails sent to people who you don't know personally. Crafting a personalized cold email can be time-consuming, but AI can significantly simplify this task.

Say your name is Amit, and you want to send a cold email to a founder named Rohit. You want to pitch him your product, Nightfall, which is a neo-CRM. The goal is to keep the email formal, yet friendly.

Prompt: *My name is Amit. Write a formal, yet approachable, cold outreach email to Rohit, pitching our product, Nightfall, which is a neo-CRM.*

ChatGPT's response could look something like:

Subject: Introducing Nightfall: A CRM Solution to Empower Your Business Growth

Dear Rohit,

My name is Amit and I am from Nightfall, a neo-CRM that is designed to streamline customer relationship management.

I understand how essential effective customer management is, especially for growing businesses. Nightfall can help you stay organized and connected with your customers with our range of features including automated customer segmentation, personalized customer journeys, and real-time analytics.

I would love to have a conversation about how Nightfall can help your business grow. Can we schedule a call sometime this week?

Best,

Amit

Let's say you also have access to Rohit's LinkedIn profile, and you can use that information to personalize your email further.

Prompt: *LinkedIn Profile: Rohit Sharma. Developer Advocate. Previously: Co-founder Strive, Intel ISEF Alum. Write a cold outreach email to Rohit, pitching our product, Nightfall, a neo-CRM. My name is Amit. Make the email formal, yet approachable. Include relevant details from his LinkedIn profile to help pitch the product.*

Output:

Dear Rohit,

My name is Amit, and I represent Nightfall, a neo-CRM solution designed for entrepreneurs like you who understand the importance of effective customer relationships.

I noted from your LinkedIn profile that you're a co-founder of Strive and have a rich experience as a Developer Advocate. Our CRM solution is designed to be intuitive and user-friendly, allowing you to focus on growing your business.

I believe Nightfall can significantly improve your customer management process. If you're interested, I would love to schedule a call to discuss further.

Best,

Amit

The above example shows how you can utilize the power of AI to craft personalized cold emails by leveraging public information from platforms like LinkedIn.

Remember, the beauty of using AI lies in its flexibility and adaptability to different styles and tones based on your requirements. Just feed in the right prompts, and you'll get the desired output!

3. Leveraging Language Models for Education

We have already discussed how language models like ChatGPT can assist you in a variety of everyday tasks, but did you know they can also be your study buddy? Language models can be incredibly useful for breaking down complex terms, and they can also generate quiz questions to test your understanding.

Demystifying Complex Concepts

Let's take an example. Imagine you're reading a textbook about cacti, and you stumble upon this passage:

"Due to adaptations to arid conditions and the presence of **less evolved woods** in Pereskioideae, the Cactaceae is among the most intriguing families, both anatomically and morphologically."

You're perplexed by the term "less evolved woods." Searching online only yields research articles that are equally confusing. So, you turn to ChatGPT for help.

You feed it the excerpt from your book along with your question:

"Due to adaptations to arid conditions and the presence of less evolved woods in Pereskioideae, the Cactaceae is among the most intriguing families, both anatomically and morphologically. What does 'less evolved woods' mean?"

ChatGPT simplifies the concept for you:

"In this context, 'less evolved woods' refers to the more primitive or basic type of wood found in early plant evolution. These types of woods lack the specialized structures and features that later evolved in more advanced plants. In the Pereskioideae subfamily of the Cactaceae family, the wood is less evolved compared to other subfamilies, making it more akin to the wood found in non-cactus plants. The less evolved wood in Pereskioideae is believed to be an adaptation to the more humid and less arid environments these plants inhabit compared to other cacti."

Now, with the term clarified, you can continue your reading with a better understanding.

Quiz Creation for Self-Assessment

ChatGPT is not only a tool for simplifying complex ideas. It's also a fantastic resource to create quizzes for self-assessment. Let's assume you have an upcoming test on calculus. **You could simply ask:**

"Generate 5 calculus I quiz questions for me."

ChatGPT responds:

1. What is the derivative of $y = x^3$?

2. Calculate the indefinite integral of $x^3 + 5x^2 + 7x + 4$.

3. Explain the product rule for derivatives.

4. Determine the equation of the line tangent to $y = x^2$ at the point (4, 16).

5. What is the second derivative of $y = x^4$?

If you are studying a more specific subject, like marketing, you can provide the model with your class notes. Let's say these are your notes from a class on March 20, 2023:

"Marketing Class Notes:

- Explored the 4 P's of marketing: Product, Price, Place, Promotion.

- Difference between B2C and B2B: B2C targets direct consumers, B2B targets businesses.

- Different methods of market research: Quantitative and Qualitative, including surveys and focus groups.

- Pricing Strategies: Skimming, Penetration, Competition, and Value.

- Segmentation: Dividing customer base by age, income, etc.

- 4 types of segmentation: Demographic, Geographic, Psychographic, and Behavioral.

- Discussed various promotional tools: Advertising, Personal Selling, Public Relations, Direct Marketing, and Sales Promotions."

You ask ChatGPT: *"Generate 5 marketing quiz questions for me based on my notes."*

ChatGPT replies:

1. Can you list the 4 P's of marketing?

2. What distinguishes B2C from B2B?

3. Name two methods of market research.

4. Can you identify four types of customer segmentation?

5. What are five promotional tools used in marketing?

This way, ChatGPT can help you review your notes and ensure you understand the key concepts.

3. Contracts: Simplifying the Complex with AI

Have you ever been daunted by the sight of a lengthy contract full of complex legal terms and conditions? Artificial Intelligence (AI) can be of assistance!

Understanding Contracts: Unravelling Legal Jargon

Often, we're faced with legal documents, like contracts, filled with confusing language that can be difficult to understand. AI can come in handy here, offering a simplified explanation of these complex terms.

Take, for example, a Non-Disclosure Agreement (NDA), which often includes such phrases as:

> *"WHEREAS, the Contractor has or may become privy to confidential information and/or trade secrets of the Government Entity; and WHEREAS, the Government Entity desires to ensure that its confidential information and trade secrets are protected from disclosure; and WHEREAS, the Contractor agrees to maintain the confidentiality of such information pursuant to the terms and conditions of this Agreement. NOW, THEREFORE, in consideration of the mutual promises and covenants contained herein, the parties agree as follows."*

To a layman, such language can be quite perplexing. However, AI can provide a more straightforward interpretation:

> *"The terms in all caps indicate that the Contractor has access to some confidential details and trade secrets of the Government Entity. The Government Entity wants to protect this information from being disclosed. The Contractor agrees to keep this information confidential according to the agreement's terms."*

AI can also help identify potential issues within a contract. For instance, an AI might be programmed to identify unfavourable clauses within a contract. While this can't replace legal advice, it can give you a starting point for discussions with your lawyer.

Creating Contracts: Automating the Drafting Process

AI can assist not only in reading and understanding contracts but also in writing them. In some instances, AI can generate a robust legal document that can be reviewed and fine-tuned by a lawyer, saving both time and resources.

Consider this scenario: A government contractor wants to write a contract for a subcontractor doing aerospace work. The contract includes details such as a duration of 1.5 years, a total worth of $2.5 million, the requirement for top-secret clearance, and payment terms. In addition, it requires the subcontractor to sign a series of agreements like a non-disclosure agreement, a non-compete agreement, a non-solicitation agreement, and a non-circumvention agreement.

An AI can generate a contract using these details, creating a strong foundation that can then be reviewed and improved by a lawyer:

"SUBCONTRACT AGREEMENT

This Subcontract Agreement (the "Agreement") is entered into by and between [Government Contractor] ("Contractor") and [Subcontractor] ("Subcontractor") as of the date set forth below.

(Contract details continue...)"

The generated contract serves as a strong foundation and covers all necessary aspects, but it would still need a professional's review for refinement.

Note of Caution:

While AI can greatly assist in understanding and creating contracts, it's important to remember that it doesn't replace professional legal advice. Always consult with a lawyer to ensure all legal documents are sound and cater to your specific needs.

Now that we've mastered the art of prompting with ChatGPT, let's embark on an exciting new frontier: AI image generation.

In the next chapter, we'll explore leading platforms like DALL-E 2 that can manifest stunning visuals from text prompts. Just as we guided ChatGPT with carefully crafted instructions, we'll learn prompting techniques tailored to unlocking the immense creativity of visual AI. From the simplicity of text to the richness of imagery, a fascinating journey lies ahead!

CHAPTER 3: AI FOR IMAGE GENERATION AND DESIGN

N ow that we have a better understanding of the world of text generation with ChatGPT, let's move forward and explore another fascinating dimension of AI: the creation of images. If you thought bringing words to life was amazing, you'll be equally impressed by how these AI models can manifest images out of thin air.

In the AI universe, finding the ideal prompt to bring forth a perfect image is an exciting challenge. While it may seem that research in this area isn't as advanced as text prompting, don't let that impression mislead you. The quest for image generation presents unique obstacles, largely due to its subjective nature and the lack of precise metrics for evaluation. But, worry not! The community of image prompting enthusiasts has made impressive strides in this field.

As we begin this journey, bear in mind that creating images with AI, like Dall-E 2, Midjourney and Stable Diffusion, is more of an art than a science. It's about inspiring the model to express and bring to life what you envision. We will cover some basic image prompting techniques to set the groundwork.

Remember, these are only the basics to get you started.

From the written word to the visually expressive, this shift in our journey is going to be a thrilling ride. Ready to move from the simplicity of black and white text to the vibrancy of full-colour images? Let's dive in.

Dall-E 2 and Stable Diffusion:

How to access Stable Diffusion:

Step 1: On your browser, navigate to: https://dreamstudio.ai/ generate

Step 2: On the Page that appears, click on login, and create an account.

Step 3: Under the Style Menu, pick a style for your image.

Step 4: Under the prompt section, write what you want to generate an image of.

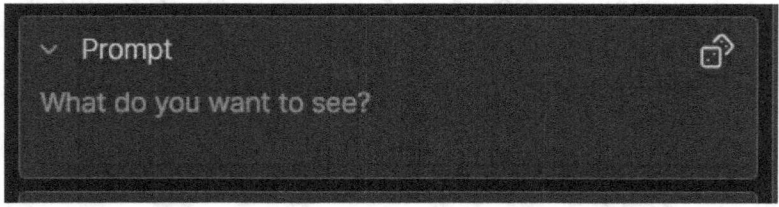

Step 5: Under the negative prompt, mention what objects your image must not have. This step is optional.

Step 6: Optionally, you can also upload your own image, and have the AI generate variations of it.

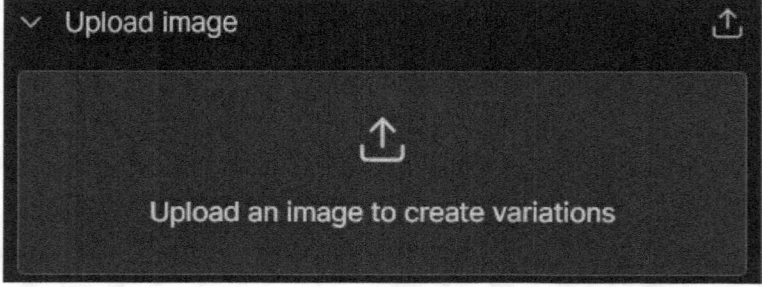

Step 7: Pick the Aspect ratio for your image, and number of variations to be created. Click on the "Dream" Button and voila! Your image will be generated.

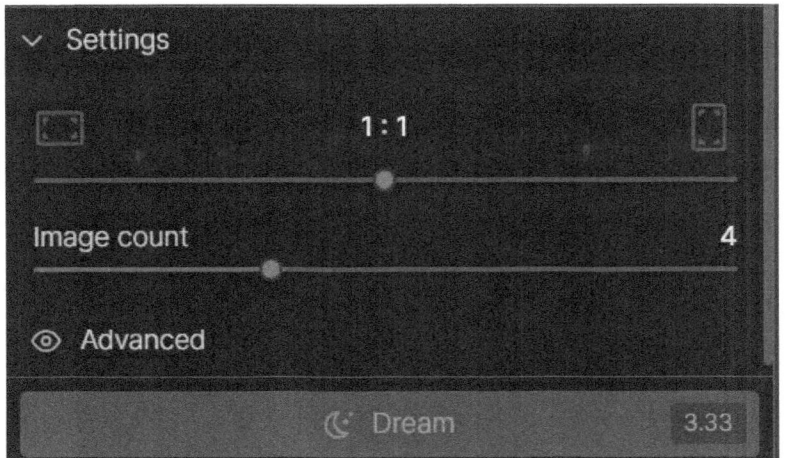

How to Access DALL-E 2:

Step 1: On your browser, navigate to: https://openai.com/dall-e-2

Step 2: Click on "Try DALL-E 2" and create an account.

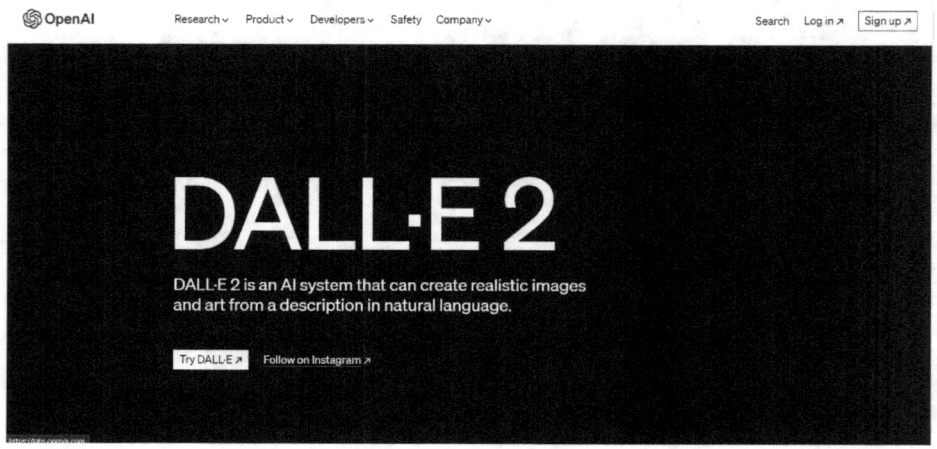

Step 3: Enter the prompt for your image, and click on "Generate".

And voila! 4 variations of your image will be generated.

Now that we've discussed how to use these fascinating tools, let's dive deeper into fine-tuning your prompts to create the most captivating images. There are certain techniques that will help you navigate this fascinating landscape more effectively.

1. Style Modifiers:

Let's start with the concept of 'Style Modifiers'. Style modifiers are like magic words that add a specific flair or theme to your creations. Imagine, you're an artist with a brush in hand and a palette full of colours. Each style modifier is a unique colour that you can add to your creation, transforming it subtly or dramatically based on your needs.

For instance, if we use terms like 'soaked in sunset hues', 'crafted from glass', or 'imagined in a video game like setting', these act as style modifiers. The beauty of these modifiers is that they can be mixed and matched to produce incredibly detailed and specific styles. They can convey information about art periods, techniques, materials, and even emulate the styles of specific artists.

To make this concept clearer, let's consider an example using DALL-E. We give the prompt 'A pyramid'. Pretty simple, right?

Edit the detailed description   Surprise me   Upload   →|

a pyramid   Generate

Now let's add our style modifiers: 'A glass pyramid, imagined in a video game like setting, soaked in sunset hues'.

Edit the detailed description   Surprise me   Upload   →|

A glass pyramid, imagined in a video game like setting, soaked in sunset hues   Generate

This single prompt with three style modifiers could generate an image of a stunning glass pyramid, glowing in the warm hues of the setting sun, set against a digital landscape.

Here's a list of some handy style modifiers you can use:

- As realistic as a photograph

- Resembling a painting

- Digital painting

- Concept art

- 3D rendering

- Cinematic lighting

- Hyper realistic

- Natural light

- Resembling film grain

As you experiment with these modifiers, you'll discover that the possibilities are truly endless. Remember, image generation with AI is not just about technology, it's a creative process. So, let's unleash your imagination and create some truly amazing art!

In our fascinating journey through the world of image generation, we've learnt about the magic of style modifiers. Now, let's turn our attention to another set of magical words - 'Quality Boosters'.

2. Quality Boosters

As the name suggests, these are words or phrases that when added to your prompts, can significantly elevate the quality of the image created.

Imagine being a chef preparing a meal. You've got all your ingredients and the dish is almost ready. But then, you add some special spices or garnishing that makes it not just good, but absolutely delectable. That's what quality boosters do to your images.

For instance, words like "spectacular", "breathtaking", and "high quality" are quality boosters that can enhance the overall aesthetic and quality of your generated images.

Let's look at an example to better understand this. Remember the pyramids we generated earlier using DALL-E with the prompt 'pyramid'? Now, let's sprinkle our quality boosters into the mix: 'A breathtaking, majestic, awe-inspiring pyramid, in high resolution'. You will notice that the resulting images are much more visually appealing and detailed!

Here's a list of several quality boosters that can add that extra 'oomph' to your creations:

- High resolution

- HD, Full HD, 4K, 8K

- Crystal clear

- Perfect lighting

- Detailed, extremely detailed

- Focused, sharp focus

- Intricate

- Aesthetically pleasing

- Hyper-realistic

- Harmonious colours

- High quality, superior quality

- Masterpiece

- Stunning

Remember, like a chef mastering the art of seasoning, the skill of using these quality boosters comes with practice. So, go ahead, experiment with them, and watch your creations come alive with extraordinary detail and quality!

Midjourney

Now, let's embark on a fascinating journey with 'Midjourney', a captivating AI image generator. Like its counterparts, Midjourney also creates images through a system of prompts. However, it's known for its unique ability to craft visually stunning and artistically refined images. This is due to its specially tailored training which enables it to produce high-quality images under specified artistic parameters.

You can interact with Midjourney through a chat interface on Discord or a user-friendly web application. The tool also plans to release an API for a more developer-friendly experience.

To use Midjourney, it's as simple as typing a command followed by your image prompt. For instance, '/imagine prompt: [IMAGE PROMPT] [--OPTIONAL PARAMETERS]'.

Midjourney also offers you additional parameters to fine-tune your images:

1. Aspect Ratio: By typing '--ar [ratio]', you can change the default aspect ratio (1:1) to a new ratio of your choice, say 16:9 or 1:2. For instance, '/imagine prompt: astronaut on a horse --ar 16:9'.

2. Chaos: The chaos parameter '--c [value]' lets you play with the unpredictability of the image. The higher the value, the more unique and unexpected the results. For instance, '/imagine prompt: astronaut on a horse --c 20'.

3. Quality: The quality parameter '--q [value]' allows you to decide how much time Midjourney should spend crafting the image, thereby affecting the overall quality. For instance, '/imagine prompt: astronaut on a horse --q .5'.

4. Seed: Using '--seed [value]', you can set a seed number which determines the initial conditions for image generation. The same seed number with the same prompt will always yield similar images. For instance, '/imagine prompt: astronaut on a horse --seed 123'.

5. Stylize: The stylize parameter '--s [value]' influences how strongly Midjourney applies its artistic flair to your image. Lower values yield images that stick closely to your prompt, while higher values create more artistically rendered images. For instance, '/imagine prompt: astronaut on a horse --s 50'.

6. Version: If you want to experiment with different versions of the Midjourney model, use '--v [version number]'. For instance, '/imagine prompt: astronaut on a horse --v 2'.

Remember, each of these parameters serves as a unique tool, helping you customize your image to your heart's content. So, go on, immerse yourself in the artistic world of Midjourney, and let your creativity flow!

CHAPTER 4: AI FOR WRITING AND CONTENT GENERATION

In the previous chapter, we tapped into the creative potential of AI image generation. Now, let's shift gears to explore AI's proficiency with another core human ability - writing. The upcoming chapter will cover innovative AI writing assistants that can help draft everything from emails to research papers. We'll learn how to move beyond generating isolated images towards creating fully-fledged stories, articles, and documents. Harnessing the dual powers of text and visuals, these versatile tools promise to augment human creativity and productivity to new heights!

Easy-Peasy AI

How to access Easy-Peasy AI:

Step 1: On your browser, navigate to: https://easy-peasy.ai

Step 2: On the page that appears, click on "Get started for free" and create an account.

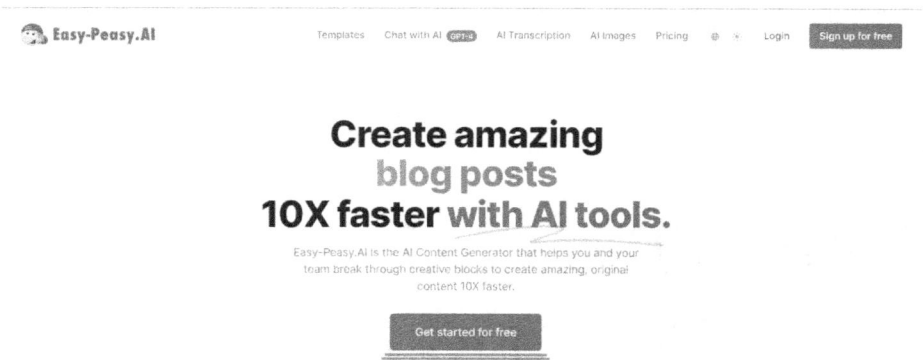

Step 3: Your Dashboard will open up, and we can now use this tool.

What does this tool do?

Let's venture into the world of 'Easy-Peasy.AI', a boon for swift and effortless content creation. This resourceful AI tool, falling under the copywriting domain, provides a myriad of templates and images to make your content creation journey smooth and delightful. Lets look at the two major use cases of this tool.

1. Templates:

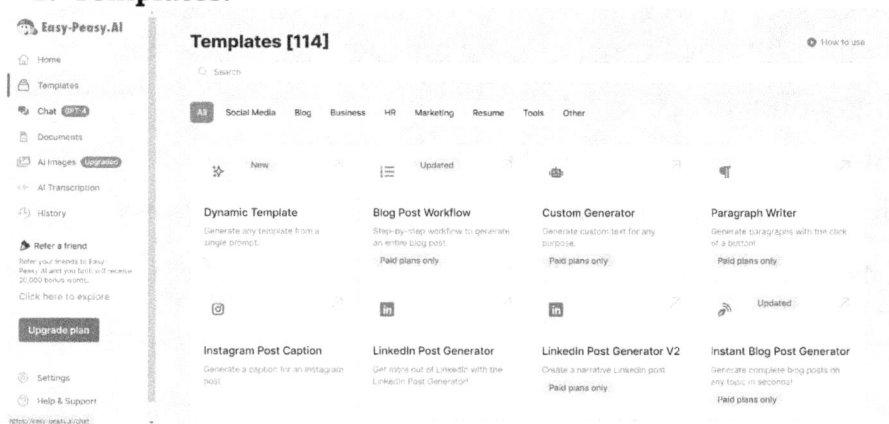

Picture this - you want to create engaging content for various platforms, be it social media, blogs, or business marketing. Easy-Peasy.AI steps in with a range of templates for all these use-cases. You simply need to fill out the fields in the templates based on your specific content requirements. For instance, consider the 'AIDA' framework template, an effective marketing blueprint that stands for Attention, Interest, Desire, and Action. This template asks for details such as your company and product name, product description, and the tone of voice you wish to adopt.

But, how do you ensure that the tool generates the best possible content? The trick lies in your input. Describe succinctly what your content is about in a sentence or two. If there are any specific details about your product or service, make sure to include them in the description field of the template. These specific facts help the AI generate content that is more aligned to your needs.

Now, don't expect the perfect output in your first run. The real fun lies in experimenting. Run your content through Easy-Peasy.AI multiple times, tweaking your inputs until you get results that truly resonate with you. You can save all the generated content in the 'History Section' on the navigation bar. Don't hesitate to pick and choose the best content from multiple runs and piece them together to craft something truly fantastic.

Remember, Easy-Peasy.AI is your playground with more than 80 templates at your disposal. Access these by clicking on the 'Templates' section in the navigation bar and let your creative instincts take over!

2. Audio Transcription:

Imagine having to transcribe a lengthy audio file - time-consuming, isn't it? Thankfully, AI transcription tools like 'Easy-Peasy.AI' are here to simplify the process. The video we're referring to illustrates how you can use such AI tools to efficiently transcribe your audio content, be it podcasts, interviews, webinars, or lectures. Not only that, these tools can help you generate succinct summaries, titles, descriptions, and even notes from your transcriptions.

The procedure is quite straightforward. Begin by uploading your audio files onto the AI transcription platform. Once the file is up, all you have to do is patiently wait for the magic to happen. The tool transcribes the entire content of your audio file, complete with timestamps.

With the transcription done, you can swiftly browse through the text instead of listening to hours of audio. This functionality is particularly beneficial for content managers or editors who might not have the luxury of time to listen to the entire audio. Extract the essence of your content by summarizing it using the AI tool.

But the usefulness of these AI transcription tools doesn't stop there. After summarizing your content, you can leverage the tool to generate appropriate titles and descriptions. All it takes is a click of the 'generate' button. Do this as many times as you need until you get a result that fits your requirements.

The beauty of these AI transcription tools, especially 'Easy-Peasy.AI', is the amount of time they save you. You stand to save hours per audio file, allowing you to focus on expanding your reach across various platforms. Once you've generated your title and description, you can use the tool's other templates to craft blog posts, social media captions, or even emails related to your audio content.

In conclusion, using AI for transcribing and summarizing your audio content is an efficient way to manage your time and resources. It assists you in quickly generating valuable components like titles, descriptions, and notes. Platforms like 'Easy-Peasy.AI' are proving to be excellent resources in making this task a breeze, regardless of whether you're dealing with podcasts or any other audio content.

Tome.app

How to Access Tome.app

Step 1: On your browser, navigate to: https://tome.app

Step 2: Click on the "Try Tome" button and create an account.

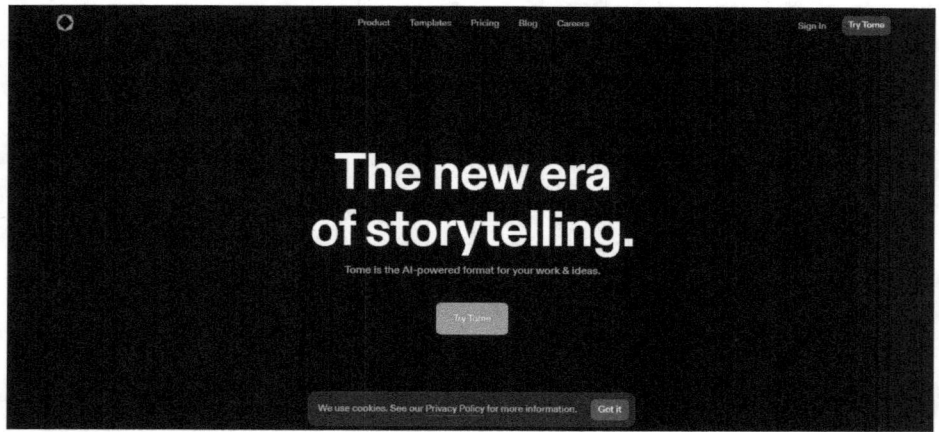

Step 3: This opens up the dashboard, and you can now use this tool.

What does this Tool Do:

Allow me to introduce you to an exciting innovation in the realm of storytelling and presentation creation - an AI-powered platform called 'Tome'. This ingenious tool is designed keeping storytellers in mind, enabling them to craft riveting narratives with the aid of advanced artificial intelligence.

Tome is not just a writing tool; it's an amalgamation of features crafted to enhance your storytelling experience. Let's look at some of these features:

1. AI-Powered Narratives: With the simplicity of a single click, Tome empowers you to generate captivating presentations, coherent outlines, or gripping stories. Your job is to provide a prompt in the command bar, and Tome takes care of the rest - creating content enriched with suitable text and illustrative images.

2. Transform Documents: Ever wished you could convert lengthy strategy documents, creative briefs, or even entire websites into something more engaging? With Tome, you can transform them into an appealing 'tome', or distil long-format content into crucial, easy-to-digest points.

3. AI-Assisted Editing: Tome elevates your content by rewriting the text to fit the tone and length you prefer, altering image styles, and adjusting image output. The aim is to bring your narrative to life.

4. Command Bar: This feature simplifies your interactions with the tool. A single command system ensures that any action you wish to perform is only a command away.

5. Interactive Content: Tome goes beyond static presentations, offering a chance to create immersive, interactive experiences. This includes adding 3D models and animations to your narrative.

6. Design Tools: Offering smart themes and responsive layouts that adapt to your content, Tome makes designing your narrative easy and intuitive.

7. Integrations: The platform seamlessly integrates with a variety of content platforms, such as Figma, Spline, Twitter, YouTube, Framer, and Miro.

8. Mobile-Responsive Layouts: Tome ensures your content is always optimally displayed, adjusting itself to fit any device.

9. Analytics: Tome lets you monitor who has viewed your 'tome'. Furthermore, the platform is planning to introduce more advanced analytics features.

The beauty of Tome is that it caters to everyone's storytelling needs. Whether you're a corporate executive, a fledgling entrepreneur, or someone creating for personal fulfilment, Tome has got you covered.

Here are some specific scenarios where Tome can be beneficial:

1. Founders & Executives: Presenting business strategies, company updates, or investor pitches.

2. Marketing & Sales: Crafting sales pitches or marketing presentations.

3. Creatives: Showcasing creative work or ideas.

4. Product & Design: Presenting product designs or prototypes.

5. Education: Crafting educational content or lessons.

6. Personal: Personal storytelling or idea presentation.

To make things even simpler, Tome offers a range of ready-to-use templates like fundraising pitches, sales pitches, team standups, company all hands, product design reviews, design portfolios, freelance pitches, and moodboards.

In essence, Tome is a complete, easy-to-use tool for creating interactive, AI-aided presentations and narratives, suitable for a wide array of professional and personal scenarios. It has the potential to revolutionize the way we create and share our stories.

To use all of these features of tome, simply click on the "Create" Button.

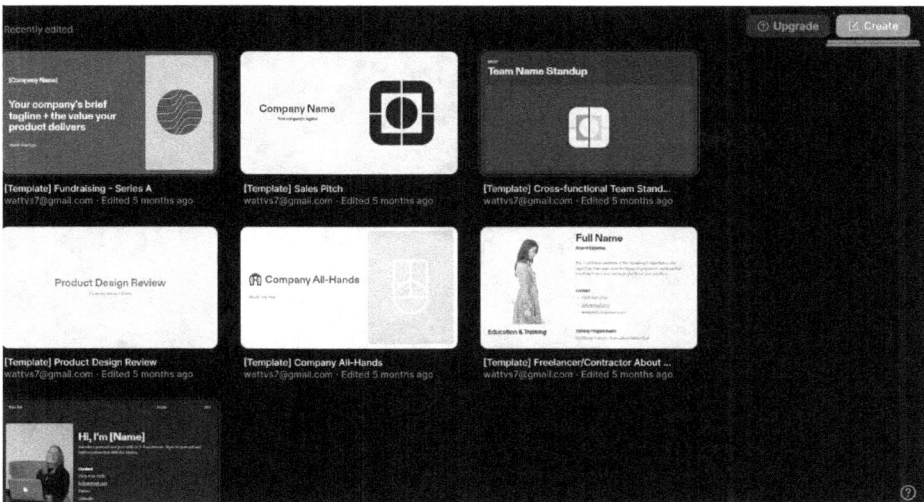

On the page that opens up, simply select what you want to create, and your prompt:

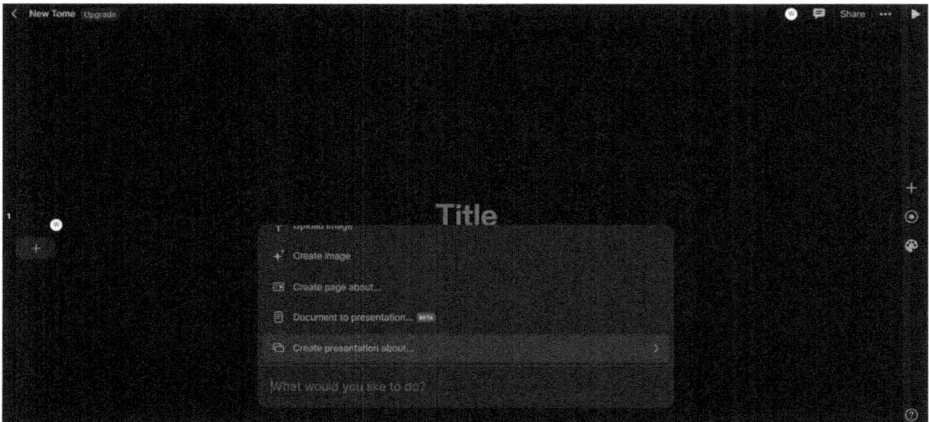

And voila! You will be presented with a beautifully created page.

Here is an example:

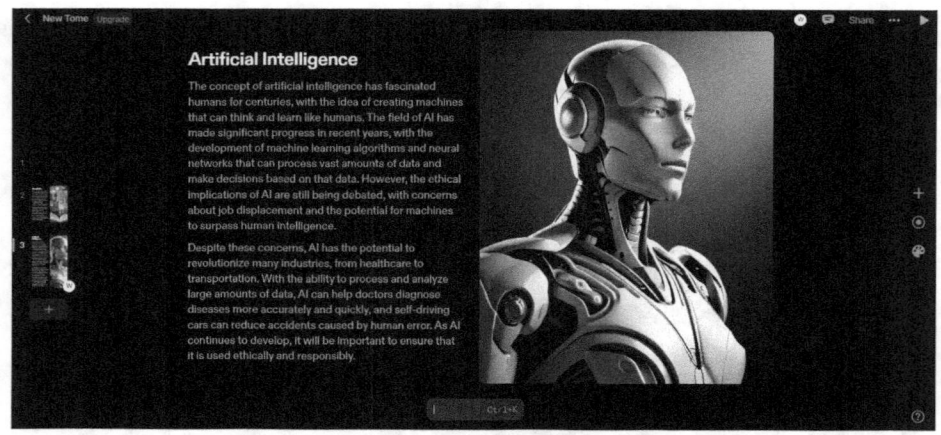

Gamma.app

How to access gamma.app:

Step 1: On your browser, navigate to this link: https://gamma.app

Step 2: Click on the "Sign up for free" button and create an account.

Step 3: This will open up a dashboard, and you can access all of its features.

Introducing 'Gamma', a brilliant AI tool under the umbrella of presentation software. To get started with Gamma, you only need to authenticate using Google or email, and you will gain access to the Gamma Tech website.

At this point, you might be asking, what exactly does Gamma offer? Well, Gamma specializes in providing a plethora of templates to aid users in creating powerful presentations and compelling websites. Allow me to introduce you to some of the highlighted templates:

Startup Pitch Deck: Tailored for startup founders or those embarking on a new venture, this template provides the scaffolding to construct an engaging narrative to secure funding.

One-Page Site - Travel Example: For those who wish to create a captivating landing page without delving into design intricacies, this template comes to the rescue. It portrays a fictional Desert Tourguide business through compelling visuals, testimonials, and clear calls to action.

Event Microsite - Tech Example 2: This template is your toolkit to create an attractive landing page for an upcoming event, serving as a comprehensive resource for attendees.

Quarterly Investor Report: Designed to deliver an easy-to-understand, interactive report for company stakeholders, this template ensures your message is well received.

Guide: Ideal for presenting your expertise or authority in a given subject area, this popular format helps convey information and present ideas online.

Six Hats Brainstorming: Based on Edward de Bono's renowned Six Thinking Hats technique, this template encourages different thinking styles or perspectives to stimulate decision-making.

Now, you must be thinking that Gamma is just about these templates. But that's just the tip of the iceberg. Gamma also includes categories for popular templates, projects & collaboration, and sales & marketing.

Let's take a look at some additional features Gamma provides:

Search: No more scrolling endlessly to find that old deck you created. Simply press Cmd/Ctrl + K and start typing the name of your deck.

Sharing and Presenting: Don't worry if your colleague or client doesn't have a Gamma account. You can share your 'Gamma' with them anyway! You can also collaboratively create a Gamma and export it as a PDF.

Editor: Need to borrow elements from one deck to another? Copy cards with ease. You can even import presentations from Google Slides, Keynote, or Powerpoint.

Integrations and Embeds: Gamma boasts a Slack integration and Google Docs import feature. It also lets you embed hundreds of apps.

Themes, Styling, Fonts: Make your deck truly yours by changing the theme or crafting a custom one. You can even upload your own custom fonts.

Now, how does one go about creating a presentation using Gamma? Here's a simple guide:

Log into Gamma and click on "New with AI" at the top of the page, next choose either "Guided" or "Text to Deck" as per your requirements.

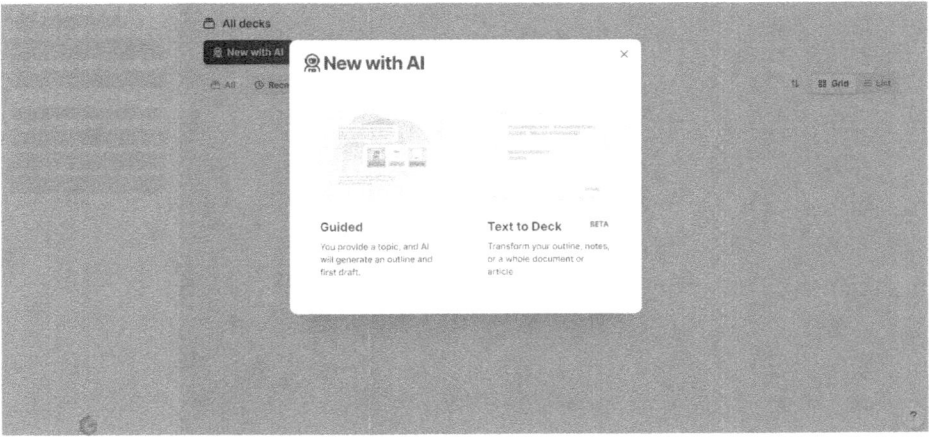

Select the kind of content you want to create - a presentation, document, or webpage. Enter your topic or keywords, select a theme, and wait for Gamma to do its magic. That's it!

Need to add more information? Easy. Just go through your workbook or textbook for ideas and add topics and vocabulary words related to your chosen theme. Once you're satisfied, review your presentation. Gamma ensures all the requested information, including vocabulary words and main topics, is included.

What about creating a webpage? The process is almost identical. Click on "New with AI," choose "Web Page," enter your topic or keywords, pick a theme, and wait. If you want to add more information, simply do so and edit the list as needed.

To sum it up, Gamma is a comprehensive tool for creating dynamic, AI-enhanced presentations, suitable for diverse professional and personal contexts. It is the epitome of user-friendly design and a testament to how AI can simplify our work processes.

CHAPTER 5: AI FOR EDUCATION

In the previous chapters, we discovered AI's potential for enhancing writing and content creation. Now, we turn our focus squarely onto the field of education. How can machines meant to replicate human intelligence aid real human learners? The upcoming chapter will uncover innovative applications of AI chatbots, virtual tutors, and more in both academic and lifelong learning settings. We'll see how AI can facilitate more interactive, accessible, and personalized education experiences. By bridging human knowledge and artificial intelligence, a new era of elevated learning awaits!

ChatPDF

How to Access:

Step 1: On Your browser, navigate to: https://www.chatpdf.com

Step 2: On your screen, you can upload your pdf, or a link that you want to use.

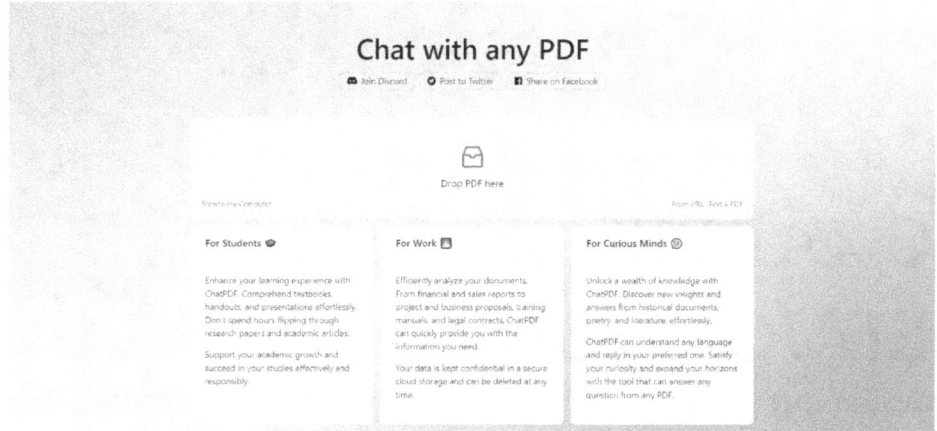

Welcome to the world of 'ChatPDF', an AI tool designed with the aim of simplifying research. As the name suggests, ChatPDF allows users to interact with PDF documents in a chat-like interface. The tool is particularly valuable in academic and educational settings, but its functionality extends beyond this scope. From students to professionals, and even to those simply eager to learn, ChatPDF has something to offer.

At the core, ChatPDF is essentially an AI-powered tool designed to help users unlock knowledge securely and efficiently. Now, you might wonder, how does it achieve this? Simply put, ChatPDF enables users to comprehend the text within a PDF document as if they were chatting with another human being. To accomplish this, ChatPDF analyses the PDF to create a semantic index of every paragraph. When a question is posed by a user, the AI then uses these indexed paragraphs to form an appropriate answer.

Now, let's delve a bit deeper into its utility and how it can revolutionize academic and educational practices.

Research: Research is a vital part of a student's academic journey, whether it's for an assignment, a term paper, or a thesis. However, it can be time-consuming and often overwhelming due to the sheer volume of available literature. This is where ChatPDF can be an academic lifesaver. It can swiftly extract precise information from large research papers or academic articles. Rather than spending countless hours perusing through these dense documents, students can ask ChatPDF specific questions about the paper's content, saving precious time and enhancing research efficiency.

Study Aid: A common challenge faced by students is grappling with complex textbooks or course materials. Comprehending these resources can be a daunting task. However, with ChatPDF, understanding complex academic content is no longer an uphill battle. By interacting with the text in a chat-like interface, students can ask questions directly from the material and receive explanations in a simplified and easy-to-understand manner. This interactive process can significantly improve comprehension and retention of information.

Group Projects: Collaborative projects are a staple in academic settings. However, ensuring everyone has the same understanding of shared documents or resources can be challenging. ChatPDF can be instrumental in such scenarios. It can assist in analyzing and discussing the contents of shared documents, fostering better understanding and facilitating effective collaboration amongst group members. Additionally, as all group members interact with the same AI interface, discrepancies in understanding are minimized, leading to a more cohesive group project.

Teaching Tool: ChatPDF isn't just useful for students, but teachers can harness its potential too. Educators can use it to convert conventional PDF documents into interactive lessons. By allowing students to ask questions directly from the lesson materials, the learning experience is not only more engaging but also fosters an active learning environment. This way, ChatPDF serves as a pedagogical tool, enhancing the teaching-learning experience in academic settings.

Language Learning: Language learning often requires a lot of reading and comprehension of text resources. With ChatPDF, interacting with foreign language texts can be less daunting and more interactive. Students can use it to ask questions directly from the text, helping them practice their language skills, and enhancing their comprehension and fluency. Moreover, as ChatPDF supports multiple languages, it becomes an effective tool for learning and practicing different languages.

One notable feature of ChatPDF is its ability to maintain data confidentiality, permitting users to delete their data at any given time. This feature ensures that while students and educators leverage AI for learning and teaching, their data remains secure and private.

Whether you are a student looking to streamline your academic research, a professional who needs to analyze reports, or simply someone with a curious mind, ChatPDF has a host of functionalities designed to simplify your interactions with text-based resources. By making dense, often intimidating documents more approachable, ChatPDF has indeed revolutionized the way we unlock knowledge.

MyAskAI

How to Access:

Step 1: On your browser, navigate to: https://myaskai.com

Step 2: Click on the "Create your AskAI" button and create an account.

Step 3: Add links and PDF files as data source for your chatbot, the answers will be based on that data.

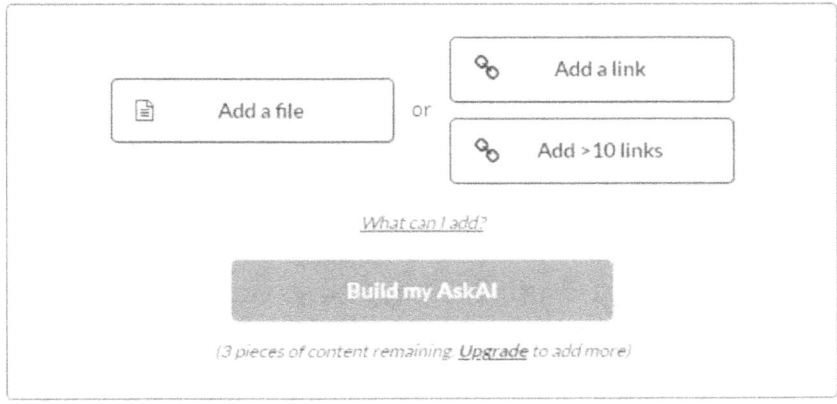

Step 4: Ask questions and check the answers

Following our discussion on ChatPDF, we now turn our attention to another advanced AI tool – 'My AskAI'. This service allows users to develop custom AI chatbots swiftly and effortlessly, with a strong emphasis on instant access to information and time-saving features.

At its core, My AskAI operates by converting text into numerical form using OpenAI's embeddings API. This process, known as 'embedding', essentially translates the complexities of human language into a format that machines can understand and analyze more effectively.

To provide a more concrete understanding, let's take a typical scenario: Suppose you have an extensive collection of information. It could be a series of lecture notes, textbooks, or research papers. Now, if you want to retrieve specific information from this vast resource, it could take hours to manually search and sift through the content. But here's where My AskAI shines.

The tool takes your content and breaks it down into smaller, manageable chunks, making it easier to handle. These chunks are then embedded using the aforementioned API, translating them into a language that the AI can comprehend. Once this embedding process is complete, these chunks are stored in a vector database, which in the case of My AskAI, is called 'Pinecone'.

With the content securely stored in the Pinecone database, the AskAI tool is now ready to tackle your queries. So when you ask a question, your query is also embedded and compared against the database. Pinecone retrieves the most similar chunks of text, and these are then passed through a customized version of ChatGPT to generate an appropriate response.

Now, one might ask, "How does My AskAI differ from ChatGPT?" The answer lies in their approach to information retrieval. Unlike ChatGPT, which might resort to creating or 'hallucinating' an answer if it doesn't have enough information, My AskAI stays truthful to the uploaded content. It will only provide answers that are contained within the provided information, ensuring accuracy and reliability.

The beauty of My AskAI is its sheer simplicity. Users need not worry about the technical intricacies, as the tool handles everything behind the scenes.

After delving into the basic functionality and several applications of My AskAI within the educational sphere, let's explore additional ways in which this innovative tool can further benefit academic settings. Here, we have compiled a range of scenarios, taking into account various levels of education, different subject areas, and diverse learning needs:

1. Homework Assistance: Everyone who's ever been a student knows that homework can be a daunting task. It requires understanding, application, and time - something that's often in short supply. With My AskAI, however, homework becomes a breeze. Let's take a history assignment as an example. Students could upload their textbooks or class notes to the tool, and then ask specific questions related to their assignment. This way, My AskAI serves as a personal, instantaneous tutor, offering precise and concise answers right when you need them.

2. Exam Preparation: As exams draw near, students are often faced with a mountain of information to review. This process can be overwhelming, but My AskAI can make it more manageable and efficient. By uploading their study materials to the tool, students can ask questions to review key concepts or clarify areas of uncertainty. Especially for open-book exams, My AskAI serves as a powerful ally, enabling students to quickly find the information they need in their notes or textbooks.

3. E-Learning Support: As digital learning environments continue to grow, so too does the need for supportive online tools. My AskAI can be seamlessly integrated into e-learning platforms, ready to provide instant answers to students' questions. This not only enriches the online learning experience but also alleviates the strain on instructors who may otherwise be flooded with queries.

4. Library Services: Libraries, as treasure troves of knowledge, can also benefit from My AskAI. The tool can be used to assist patrons in navigating digital resources, such as e-books or online databases. With a custom chatbot capable of answering queries about the library's resources, patrons can quickly and easily find the information they need.

5. Research Projects: For students undertaking complex research projects, My AskAI serves as an ideal companion. Researchers can feed the tool with documents, scholarly articles, or reports, and then pose questions to extract key information or insights. This process streamlines research, making it easier to sift through copious amounts of text data and retrieve relevant information.

6. Accessibility: Education should be accessible to all, and My AskAI can play a crucial role in making this possible. For students with visual impairments or reading difficulties, the tool can read out text from PDF documents or answer queries verbally. This feature ensures that all learners can access and engage with learning materials, regardless of their circumstances.

7. Tutoring Services: Tutors can harness the power of My

AskAI to extend their services. By creating a chatbot that can answer common questions or explain complex concepts, students can receive help at any time, even when a human tutor isn't available. This function provides consistent support to students and enhances the effectiveness of tutoring services.

8. Administrative Tasks: Beyond its educational applications, My AskAI can streamline administrative tasks within schools. For example, the tool can answer frequently asked questions about school policies, procedures, or events. This automation saves time, improves efficiency, and ensures that everyone has access to up-to-date, accurate information.

9. Language Learning: As My AskAI supports a wide range of languages, it's a perfect fit for foreign language learning. Students can upload a document in the language they're learning and use the tool to ask questions about the text. This interaction supports reading comprehension and provides an interactive way to practice a new language.

10. Literature Analysis: In literature classes, students can use My AskAI to delve deeper into novels, plays, or poems. By asking the tool about plot elements, characters, themes, or literary devices, students can gain a more thorough understanding of the text. This tool, therefore, serves as a personal literature guide, always ready to shed light on the intricacies of any literary work.

In essence, My AskAI is more than just an AI chatbot. It's a multifaceted educational tool that caters to a wide range of needs and learning environments. From helping students with their homework and exam preparation to assisting in research projects and language learning, My AskAI can be an indispensable companion in the journey of education. Its seamless integration into various educational settings makes it a tool worthy of consideration for any academic institution or learner.

CHAPTER 6: AI FOR SEO AND BUSINESS

So far, we've explored uses of AI in creative domains like writing, design, and multimedia. But AI's utility extends well beyond the arts - it has invaluable business applications too. Coming up, we'll learn how AI can elevate marketing, streamline workflows, and even optimize websites through SEO. Whether you're an executive, marketer, or entrepreneur, AI is ready to augment your work. Let's dive in to realize how AI can catalyse business growth and online visibility!

Rytr:

How to Access:

Navigate to https://rytr.me , click on the "Start Ryting" Button and create an account to start using this tool.

What does this tool do?

As we move forward in our exploration of AI writing tools, our next stop is "Rytr". Rytr is an AI-powered assistant that uses sophisticated machine learning algorithms to generate high-quality content in a matter of seconds. Primarily recognized as a "copywriting" tool, it's a Swiss Army knife in the world of content creation, offering help for a wide array of tasks. In this section, we will dissect Rytr's features, its performance, its implications in an educational setting, and its usefulness for SEO tasks.

Rytr: The Essentials

Rytr stands out in the AI writing tools landscape thanks to its user-friendly interface and vast range of features. It's capable of generating content for over 40 different use-cases, such as drafting business idea pitches, crafting social media ads, brainstorming blog ideas, and even laying down the groundwork for writing projects with frameworks like AIDA.

This adaptability, coupled with its advanced text editing options, allows Rytr to serve users in numerous content creation tasks, and makes it a powerful ally for writers.

One of Rytr's major strengths is its multi-language support.

Capable of generating content in 29 languages, including Hindi, Rytr ensures that your writing needs are met, no matter the language. In addition, Rytr offers a convenient browser extension, enabling you to conjure up engaging copy while working on an email, social post, or blog post.

When it comes to the quality of the content it produces, Rytr presents a mixed bag. While it can generate commendable short snippets of copy, the quality tends to fluctuate with longer pieces. This means that users might need to invest some time in editing and fact-checking to ensure the content aligns with their standards. However, given the right guidance, Rytr can generate respectable ideas for blogs, product descriptions, and even engaging video intros. In the realm of academia, Rytr proves to be an invaluable ally for students and teachers alike. Students can leverage Rytr's capabilities to generate essay ideas, create comprehensive outlines for research papers, and draft professional emails to their professors. The process is straightforward, allowing students to focus on the learning experience rather than getting bogged down by the tedious aspects of writing.

On the other side of the classroom, educators can harness Rytr to simplify their tasks. They can create detailed lesson plans, draft informative emails to their students, and even generate academic blogs to share their insights with a wider audience. This ability to automate routine writing tasks gives teachers more time to focus on their primary responsibility – educating and inspiring young minds.

Rytr and SEO

Search Engine Optimization (SEO) is crucial for anyone looking to improve their website's visibility on search engines. Rytr's advanced algorithms can help craft SEO-friendly content, from generating keyword-rich blog ideas to creating meta-descriptions that capture the essence of your content. By providing insightful SEO suggestions, Rytr aids in boosting your site's ranking and increasing organic traffic.

In summary, Rytr serves as an effective AI writing assistant, offering diverse features and multi-language support. While it requires some degree of user editing to refine its AI-generated content, it shines in generating short-form content and providing assistance in multiple languages. It presents itself as a valuable tool in academic contexts, facilitating students and educators in their writing tasks. With its SEO-centric capabilities, Rytr is a strong contender in the landscape of AI writing tools.

CHAPTER 7: AI FOR VOICE GENERATION

In the previous chapters, we investigated uses of AI for creativity, education, and business. Now, we'll explore how AI can lend its voice - quite literally - to augment human communication and storytelling. From voice cloning to speech synthesis, the upcoming chapter will showcase AI's remarkable ability to generate realistic and customizable vocalizations from text. We'll learn how these tools can increase accessibility, facilitate language learning, and open new creative possibilities. Let's dive in to hear AI's perspective!

What is Voice AI?

Voice AI refers to AI systems designed specifically for speech recognition, voice synthesis, and vocal manipulation. Here are some key voice AI categories:

- Speech Recognition: Converts speech to text by analyzing acoustic signals. Used in voice assistants like Siri.

- Speech Synthesis: The reverse of speech recognition. Converts text into human-like speech using computer voices. Used for navigation guidance, audiobooks, etc.

- Voice Cloning: Mimics the voice of an existing person by studying their speech patterns. Useful for personalization.

- Voice Transformation: Alters pre-recorded voices by

modifying pitch, tone, accents and other qualities. Enables vocal experimentation.

Voice AI leverages deep learning and vast datasets of human speech to deliver realistic vocal results. The criteria for judging voice AI include naturalness, clarity, diversity of expression, and accuracy of imitation. As these technologies continue advancing, the applications are vast.

Lets look at some tools to leverage this technology:

ElevenLabs

How to Access:

Step 1: On your browser, navigate to: https://beta.elevenlabs.io/

Step 2: Sign up for an account

Step 3: This brings you to the **Speech Synthesis** page, where you will have the options to select a voice, modify voice settings, select a language and input your text.

Speech Synthesis

Settings   Rachel   ⌄   + Add voice

            Voice Settings   ⌄

            Eleven Monolingual v1   ⌄

Text   Type or paste text here. The model works best on longer fragments.

What does this tool do?

Let's delve into Eleven Labs, an advanced AI-based platform specializing in voice synthesis and voice cloning. This remarkable platform offers a wide range of features that enable users to create, customize, and generate high-quality voiceovers with ease. In the following sections, we will explore how to get started with Eleven Labs, speech synthesis, available models, and prompting techniques.

Getting Started with Eleven Labs

Signing up for Eleven Labs is a breeze. Users can choose to sign up using their Google or Facebook accounts or simply use their email. Once signed up, users are directed to the Speech Synthesis page, where they can immediately begin using the service. By typing text into the designated box and clicking "generate," the platform vocalizes the written words, bringing them to life through synthesized voices.

Voice selection:

The first row of settings is the voice selection. This is where you will select all of the pre-made voices. There's a small black arrow next to each voice that you can click to preview what the voice sounds like without having to use up characters. Keep in mind that you need to have generated something prior to having it show up on new voices.

Once a voice has been selected, users can proceed to the Speech Synthesis tab to generate voiceovers. It's important to note that the AI settings are nondeterministic, meaning that each generation may yield slightly different results. The Voice Settings significantly impact the voice's sound and performance.

Voice Settings:

1. Stability:

The stability slider determines how stable the voice is and the randomness of each new generation. Lowering this slider introduces a broader emotional range for the character - this, as mentioned before, is also influenced heavily by the original voice. Setting the slider too low may result in odd performances that are overly random and cause the character to speak too quickly. On the other hand, setting it too high can lead to a monotonous voice with limited emotion.

2. Similarity

The similarity slider dictates how closely the AI should adhere to the original voice when attempting to replicate it. If the original audio is of poor quality and the similarity slider is set too high, the AI may reproduce artefacts or background noise when trying to mimic the voice if those were present in the original recording.

Users often set the stability around 40 and similarity around 75 as starting points, adjusting as needed through multiple generations until they achieve the desired performance.

Available Models: Monolingual and Multilingual

Eleven Labs offers two models: monolingual and multilingual.

The monolingual model was trained on an English dataset, while the multilingual model incorporates data from several languages, including French, German, Hindi, Italian, Polish, Portuguese, and Spanish. The multilingual model is still in its experimental stage and may have some quirks that are being addressed. For primarily English voiceovers, it is recommended to use the monolingual model, while keeping an eye on the multilingual model's progress.

Prompting: Control and Expressiveness

Users have control over the pacing, emotion, and pauses in their generated voiceovers. To introduce pauses, users can use a simple dash (-) or an em-dash (—). Additionally, ellipsis (...) can be used to add pauses with a hint of "hesitation" or "nervousness" to the voice. Inserting line breaks can also prompt the AI to recognize changes in the text and adapt accordingly, resulting in natural-sounding pauses.

For expressing specific emotions, users can write in a style similar to that of a book. Dialogue tags, such as "he said, confused," or "he shouted angrily," can guide the AI in understanding the desired emotional tone. Users can experiment with different prompts to create highly customized voiceovers that accurately reflect the intended emotion.

In Summary, Eleven Labs provides an all-encompassing platform for creating and customizing high-quality voiceovers. With its user-friendly interface and precise control over voice characteristics, users can tailor their voiceovers to meet their specific needs. Whether it's for educational purposes, multimedia projects, or any other application that requires voice synthesis, Eleven Labs offers a powerful and versatile solution.

Play.ht

How to access:

On your browser page, navigate to this link: https://play.ht, and sign up for a new account.

What does this tool do:

This tool works similar to ElevenLabs, and you can generate voiceovers in various voices using text to speech. However, there is one additional feature that we will look into: **Voice Cloning.**

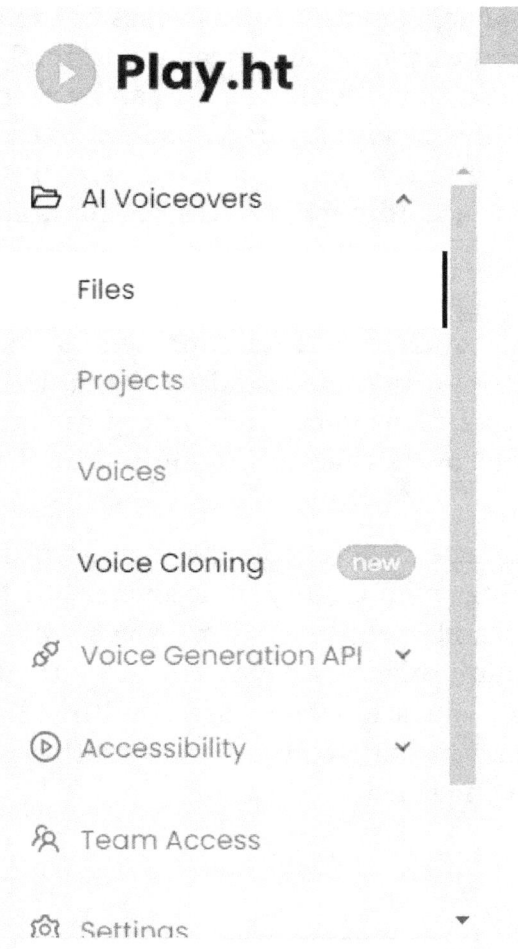

Using this feature, you can clone an existing person's voice, by providing the tool with recording of this person's voice.

On the navigation bar, click on **Voice Cloning** as shown in the picture.

Next click on the "Clone a voice" button.

You can pick from the below two options:

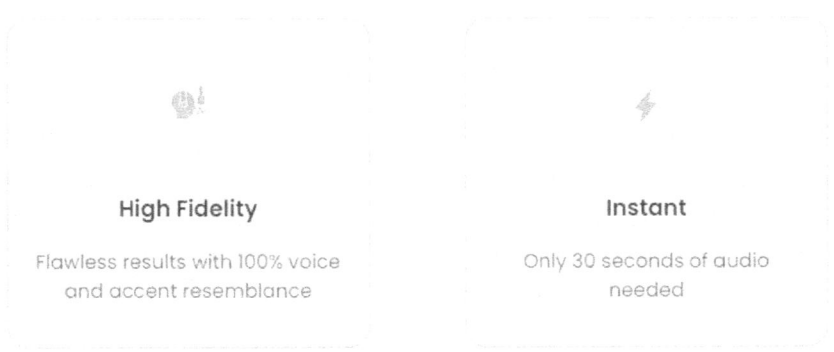

Then, just upload your audio file, wait for it to process and you can then make this cloned voice say anything with text to speech!

CHAPTER 8: AI ETHICS AND RESPONSIBLE USE

Introduction to AI Ethics

Artificial intelligence, a marvel of modern technology, holds the power to transform our world. As it permeates every sphere of life, from healthcare and education to business and entertainment, it brings a host of benefits—efficiency, precision, and the ability to discern patterns and insights beyond human reach. However, with its immense capabilities comes an equally great responsibility: to ensure that AI's impact on society aligns with our moral principles and ethical standards.

The Importance of Ethics in Technology and AI

The interplay of ethics and technology is not a new phenomenon. Throughout history, every significant technological innovation has prompted us to revisit our ethical boundaries and reassess our moral norms. From the invention of the printing press, which spurred debates about intellectual property and freedom of information, to the advent of genetic engineering, which sparked discussions about the sanctity of life and the limits of scientific intervention, technology has consistently nudged us to redefine our ethical landscape.

In the realm of AI, this ethical reevaluation takes on a heightened urgency. Unlike many of its technological predecessors, AI has the capacity to make decisions, to learn from experience, and to interact in ways that mimic human intelligence. As AI systems become increasingly integrated into our lives and societies, the decisions they make and the actions they take inevitably intersect with human values, norms, and ethics. They influence our lives in profound ways, from shaping our social media feeds to diagnosing medical conditions, approving loans, and even driving our cars.

This is why ethics is of paramount importance in AI. The algorithms that sift through our data, that make decisions on our behalf, and that silently shape our digital experiences must do so in a way that is fair, transparent, and respectful of our rights and values. They must not simply serve us; they must do so in a way that aligns with our ethical principles and societal norms.

The Ethical Dimensions of AI

When we speak of ethics in AI, we are referring to a broad and multifaceted domain. There are several key dimensions to consider:

- **Fairness:** AI systems must be fair in their decisions and actions. They must not favor one group over another based on characteristics such as race, gender, or socio-economic status. This requires careful attention to how AI systems are trained and how they make decisions.

- **Justice:** AI systems must be just in their impact. They should not contribute to societal inequalities or harm certain groups disproportionately. This calls for a systemic perspective, considering not just individual AI systems but also how AI as a whole influences society.

- **Transparency:** AI systems should be transparent in their workings. People should be able to understand how an AI system makes a decision and on what basis. This is key to building trust in AI systems and ensuring that they can be held accountable.

- **Privacy:** AI systems must respect privacy. They should not intrude into people's private lives or collect data without consent. This involves careful data management and the use of privacy-preserving technologies.

- **Autonomy:** AI systems should respect human autonomy. They should not manipulate people or make decisions on their behalf without their informed consent.

In the following sections, we will explore each of these dimensions in greater detail, delving into the complexities of bias, privacy, transparency, and accountability in AI. We will look at real-world examples, discuss the challenges and dilemmas, and explore possible solutions. As we navigate this ethical landscape, our guiding principle will be clear: the goal of AI is not just to serve, but to serve ethically.

Bias in AI

Artificial intelligence, a mirror reflecting the world it learns from, is susceptible to the same biases that permeate our societies and cultures. Bias in AI, an echo of human prejudice, manifests as a systematic error that skews the decisions and actions of AI systems. It is a pervasive issue, insinuating itself into various aspects of AI, from the data used for training to the algorithms that process this data.

Understanding Bias

Bias in AI can be broadly categorized into three types: pre-existing, technical, and emergent.

Pre-existing bias refers to biases that exist in society and are reflected in the data used to train AI systems. If the training data are skewed towards a particular group, the AI system is likely to perform better for that group and may inadvertently discriminate against others.

Technical bias arises from the limitations or characteristics of the AI system itself or the data processing techniques employed. For example, an AI system designed to process text data might struggle with languages that are not well-represented in its training data, leading to poorer performance for speakers of those languages.

Emergent bias surfaces when AI systems are deployed in the real world. Even if an AI system has been carefully designed and trained to be unbiased, biases can emerge based on how the system is used or the context in which it operates.

Case Studies of Bias in AI

One of the most striking examples of AI bias is found in the realm of facial recognition technology. Studies have shown that these systems exhibit racial and gender bias, performing less accurately for individuals with darker skin tones and for women. This bias, a reflection of the underrepresentation of these groups in the training data, raises serious ethical concerns. When such biased systems are used in critical areas like law enforcement or job recruitment, they can reinforce existing inequalities and unfairly disadvantage certain groups.

Another example comes from the world of natural language processing (NLP), where AI models trained on large text corpora have been found to reproduce and amplify gender stereotypes present in the data. For instance, an AI language model might associate men with careers and women with family roles, reflecting biases in the training data. This is a concern for applications like automated resume screening, where such biases could lead to unfair outcomes.

Mitigating Bias

Mitigating bias in AI is a challenge that requires concerted effort across multiple fronts. It starts with the recognition that bias is a problem and a commitment to address it.

One crucial step is to ensure that the data used to train AI systems are representative of the diverse range of individuals and groups that the systems will serve. This requires careful data collection and curation, as well as ongoing monitoring to detect and correct any biases that may emerge.

Algorithms can also be designed to be more robust to bias. Techniques like fairness-aware machine learning aim to incorporate fairness considerations directly into the AI training process, by adjusting the learning algorithm to reduce bias in its predictions.

Bias audits are another important tool. These involve testing AI systems to detect biases in their performance, using a variety of metrics and methods. Bias audits should be conducted regularly and the results used to improve the system.

In addition, the teams that design and deploy AI systems should themselves be diverse, bringing a wide range of perspectives and reducing the risk of unintentional bias. Diversity in AI development is not just a matter of fairness—it is a matter of quality and effectiveness.

Addressing bias in AI is not a one-time task, but an ongoing commitment. It requires vigilance, openness to feedback, and a willingness to iterate and improve. As we continue to harness the power of AI, let us strive to ensure that this power is wielded fairly and equitably. In the next section, we turn our attention to another critical aspect of AI ethics: privacy.

Privacy and AI

As we tread deeper into the realm of artificial intelligence, we are met with an intricate dance between utility and privacy. AI systems, in their quest for optimization, learning, and prediction, thrive on data. However, this voracious appetite for data brings with it a pressing concern: the risk of infringing upon individual privacy.

AI and Data Collection

Data forms the lifeblood of AI systems, especially those employing machine learning. They learn from data, much like a child learns from observing the world. The more data they have, the more patterns they can discern, the better their predictions and decisions. However, this creates a tension between the utility of AI systems and the privacy of individuals.

The data that feed AI systems often include personal information about individuals—what they buy, where they go, who they interact with, even what their heart rate was during their morning run. In the wrong hands, or used in the wrong way, such data could be used to infringe on individual privacy, leading to outcomes ranging from targeted advertising and price discrimination to identity theft and surveillance.

Privacy Breaches and AI

The potential for privacy breaches is not just a theoretical concern. There have been several instances where AI systems, directly or indirectly, led to breaches of privacy.

For instance, consider an AI system designed to recommend health products based on a user's online activity. If not properly designed and regulated, such a system could end up revealing sensitive health information about the user. This could happen if the system starts recommending certain products to a user in a public setting (like a shared computer), thereby disclosing the user's health conditions to others.

In another case, a large technology company was found to be using AI to analyze emails and serve targeted advertisements, raising concerns about user privacy. The company eventually stopped this practice, but the incident sparked a broader discussion about the extent to which AI systems should be allowed to analyze personal data for commercial purposes.

Protecting Privacy

Protecting privacy in the age of AI is a multifaceted challenge that requires a combination of technical, legal, and ethical solutions.

On the technical front, there are several strategies to safeguard privacy while leveraging AI. One approach is to use anonymization techniques to remove or obfuscate identifying information from data. For instance, a dataset containing medical records could be anonymized by removing names, addresses, and other personal identifiers, so that the remaining data cannot be linked back to individuals.

Another approach is to use privacy-preserving AI algorithms. Techniques like differential privacy add a carefully calibrated amount of noise to the data or the AI system's outputs, ensuring that the results cannot be used to infer sensitive information about individuals. Federated learning, another technique, allows AI systems to learn from data located on many different devices, without the data ever leaving those devices.

On the legal and ethical fronts, robust data governance policies are crucial. These policies should specify what data can be collected, how it can be used, and how long it can be stored. They should also ensure that individuals are informed about the data collection and have the opportunity to give their consent.

Privacy is not just a technical issue—it is a fundamental human right. As we develop and deploy AI systems, we must remain vigilant in protecting this right. By combining technical ingenuity with ethical foresight, we can harness the power of AI while respecting and safeguarding individual privacy. In the next section, we will explore the theme of transparency in AI, a cornerstone of ethical AI practices.

Transparency and Explainability in AI

As the tendrils of AI reach deeper into our lives, making decisions and taking actions that directly impact us, the need for transparency and explainability in AI systems becomes increasingly paramount. We are entitled to understand the mechanisms that operate behind the scenes of these influential machines. To trust the unseen intellect of an AI system, we must pull back the curtain on its decision-making process, shedding light on the once opaque operations.

The Black Box Problem

At the heart of the demand for transparency lies a notorious problem in the field of AI, often referred to as the "black box" problem. The black box metaphor encapsulates the idea that while we can observe the inputs and outputs of an AI system, the internal workings—the process through which the AI system transforms inputs to outputs—remains hidden from view.

This issue is particularly prominent in certain types of AI systems, such as deep learning models. These models, composed of multiple layers of interconnected nodes (or "neurons"), can learn highly complex patterns and make highly accurate predictions. However, understanding how they arrive at a particular decision can be a daunting task, akin to trying to understand the thought process of a human brain based solely on its neural activity.

The black box problem is not just an academic curiosity. It has real-world implications. For instance, if an AI system denies a loan application or a job application, the person affected has a right to know why they were denied. If the AI system is a black box, providing a satisfactory explanation can be challenging.

Case Studies

The need for transparency and explainability in AI has been highlighted in several real-world cases.

For example, consider the case of a teacher who was dismissed based on scores from an AI system used to evaluate teacher performance. The system used complex statistical models to predict student scores and rank teachers based on these predictions. However, the teacher was not given a clear explanation of why her score was low, and the system's developers admitted that the model was difficult to explain. The teacher sued the school district, arguing that she had a right to understand the reasons for her dismissal.

In another case, a healthcare provider used an AI system to predict which patients would benefit from a care management program. However, the system was biased against black patients, ranking them as less needy than equally sick white patients. The bias was traced back to the system's use of healthcare costs as a proxy for health needs. Because black patients incurred lower costs than white patients with the same level of sickness (due to systemic disparities in healthcare access and utilization), the system incorrectly inferred that they were healthier. Without transparency into the system's decision-making process, this bias went unnoticed for several years.

Making AI Transparent

Despite the challenges, several techniques have been developed to make AI systems more transparent and explainable.

One approach is to use inherently interpretable models, such as decision trees or linear regression, which provide clear explanations for their decisions. However, these models are often less powerful than their black-box counterparts and may not be suitable for complex tasks.

Another approach is to use post-hoc explanation techniques, which aim to explain the decisions of a trained AI model after the fact. One popular technique is LIME (Local Interpretable Model-Agnostic Explanations), which explains an AI system's decisions by approximating its behavior with a simpler, interpretable model. Another technique is SHAP (SHapley Additive exPlanations), which assigns each feature an importance value for a particular prediction, based on concepts from cooperative game theory.

While these techniques provide valuable insights, they are not without limitations. For instance, they may not always be reliable, and their explanations may not capture all the nuances of the AI system's decision-making process.

The quest for transparency and explainability in AI is ongoing. It is a challenging journey, but one that we must undertake. As AI systems become more integral to our lives, we must strive to make them not just intelligent, but also transparent and accountable. In the next section, we will delve into the intricacies of accountability in AI, a critical facet of AI ethics.

Accountability in AI

As we entrust artificial intelligence with an ever-expanding range of decisions and tasks, the need for accountability grows in tandem. When an AI system makes a mistake, causes harm, or behaves unexpectedly, it is vital to have mechanisms in place to hold the relevant parties accountable. But who is to blame when an AI errs? Is it the developers who programmed it, the users who deployed it, or the AI itself? The question of accountability in AI is complex and multifaceted, involving considerations from law, ethics, and computer science.

The Blame Game

Accountability in AI is not a straightforward matter. AI systems are complex artifacts, involving a diverse array of components and stakeholders. They include not just the AI algorithms themselves, but also the data used to train them, the hardware on which they run, and the human users who interact with them. Each of these components is associated with different individuals or groups—the data scientists who design the algorithms, the engineers who collect and curate the data, the companies that manufacture the hardware, and the end-users who deploy the AI system. When an AI system makes a mistake, determining who is at fault can be like untangling a web.

The question of AI accountability is further complicated by the autonomy and learning capabilities of AI systems. Unlike traditional software, which follows predefined rules, AI systems can learn from experience, adapt to new situations, and make decisions based on complex, often opaque, computations. This makes it difficult to predict their behavior and to trace the cause of a particular action or decision. In some cases, it might even be argued that the AI system itself is responsible, raising the controversial question of whether AI systems can or should bear legal or moral responsibility.

AI and the Law

Current legal frameworks are often ill-equipped to handle the complexities of AI accountability. Laws and regulations, traditionally based on human actors and predictable software, struggle to accommodate the autonomy and unpredictability of AI systems.

In response, legal scholars and policymakers have proposed various approaches to reforming the law to better accommodate AI. Some have suggested adapting existing legal doctrines to hold the creators or users of AI systems responsible. For example, product liability law, which holds manufacturers responsible for defects in their products, could be extended to cover errors or harms caused by AI systems.

Others have proposed more radical changes, such as creating a new legal status for AI systems or instituting a mandatory insurance scheme for AI-related harms. These proposals are still at the stage of academic debate, but they highlight the need for legal innovation in response to the challenges posed by AI.

Ensuring Accountability

Beyond the realm of law, there are several technical and organizational measures that can help ensure accountability in AI.

On the technical side, robust testing and validation of AI systems can help detect and correct errors before deployment. Techniques like adversarial testing, where the AI system is subjected to challenging inputs designed to cause errors, can be particularly effective.

Transparency and explainability, discussed in the previous section, also play a crucial role in accountability. If an AI system's decision-making process is understandable, it is easier to determine why a mistake occurred and who is responsible.

On the organizational side, accountability can be promoted through internal audits, third-party audits, and the establishment of clear lines of responsibility for AI systems. Companies and institutions can also foster a culture of accountability, emphasizing ethical design and responsible use of AI.

Conclusion

The question of accountability in AI is a challenging yet crucial aspect of AI ethics. As we increasingly delegate decisions and tasks to AI systems, we must ensure that these systems and their human overseers are held accountable for their actions. This will require not only legal and regulatory innovation, but also technical measures and a strong commitment to ethics at all levels of AI development and use.

In the next chapter, we will explore the exciting and complex interplay between AI and the future of jobs—a topic that touches on the hopes and fears of millions as we stand on the brink of the AI revolution.

CHAPTER 9:
THE FUTURE OF
JOBS AND AI

As we stand on the precipice of the Fourth Industrial Revolution, artificial intelligence (AI) is at the helm, steering us towards a future interwoven with digital, physical, and biological threads. Much like the revolutions that preceded it—each with its distinct character, from the mechanization of the First Industrial Revolution to the digitalization of the Third—the Fourth Industrial Revolution carries profound implications for the world of work. At the heart of these transformations is AI, an engine of change that is simultaneously awe-inspiring and anxiety-inducing. As AI continues to evolve, so too does its impact on jobs, prompting a range of responses, from optimistic anticipation to concern and fear. In this chapter, we will navigate this complex terrain, exploring the multifaceted interplay between AI and jobs.

Part I: The Dual Faces of AI: Automation and Augmentation

AI, in its interaction with the job market, wears two distinct faces: one of automation and one of augmentation. These twin aspects of AI, often seen as opposing forces, are in fact two sides of the same coin, each with its unique implications for the future of work.

Understanding Automation

Automation—the process of using machines or technology to perform tasks that were once carried out by humans—is not a new phenomenon. It has been a defining feature of industrialization, from the mechanized looms of the 19th century to the assembly lines of the early 20th century. However, AI brings a new dimension to automation. With its ability to learn from data, adapt to new situations, and make complex decisions, AI has the potential to automate not just manual and repetitive tasks, but also cognitive tasks that require understanding, judgment, or creativity.

The impact of automation on jobs is a complex issue, with both positive and negative aspects. On the positive side, automation can lead to increased productivity, cost savings, and improved quality. By taking over routine tasks, it can free up human workers to focus on more complex and creative tasks, potentially leading to more fulfilling jobs.

On the downside, however, automation also raises concerns about job displacement. As AI systems become more capable, there's a fear that they could replace human workers in a wide range of jobs, leading to job losses and increased inequality. While some jobs are more vulnerable to automation than others— typically those involving routine, predictable tasks—no job may be completely immune.

Consider the example of self-driving cars and trucks. With advances in AI and related technologies, we're moving closer to a future where vehicles can drive themselves, without a human driver. This has enormous potential benefits, from reducing traffic accidents to improving fuel efficiency. However, it also poses a threat to jobs in driving-related occupations, from truck drivers to taxi drivers.

Understanding Augmentation

In contrast to the narrative of automation and job displacement, there's another, more optimistic narrative about AI and jobs: augmentation. This is the idea that AI can be used to enhance human capabilities, helping humans to perform their jobs better, rather than replacing them.

AI augmentation can take many forms. It can involve automating the routine parts of a job, leaving the human worker free to focus on the more complex and creative parts. It can involve providing the worker with insights or recommendations, based on AI's ability to analyze large amounts of data and detect patterns. Or it can involve creating new capabilities that were not possible before, enabling entirely new ways of working.

For example, in healthcare, AI is being used to augment the capabilities of doctors and other healthcare professionals. AI systems can analyze medical images, like X-rays or MRI scans, and highlight areas that may indicate a disease, helping doctors to make more accurate diagnoses. They can analyze a patient's medical history and lifestyle data, and provide personalized recommendations for treatment or prevention. By augmenting the capabilities of healthcare professionals, AI has the potential to improve the quality of care, reduce errors, and make healthcare more personalized and proactive.

However, augmentation also carries its own challenges. It often requires workers to learn new skills and adapt to new ways of working, which can be difficult, especially for those who are already well-established in their careers. It can also lead to concerns about privacy, trust, and the dehumanization of work, especially when AI systems are used in sensitive areas like healthcare or education.

In the next section, we will delve deeper into the implications of AI-driven automation and augmentation, exploring their impact on the job market and what they mean for the future of work.

Part II: The Skills Revolution: Reskilling and Upskilling

As the landscape of work is reshaped by AI, it becomes increasingly clear that the workforce must adapt to keep pace with these changes. A crucial part of this adaptation involves reskilling and upskilling, two strategies that hold immense potential for preparing individuals for the future of work.

The Importance of Reskilling

Reskilling, or the process of learning new skills to transition into a new job or industry, is becoming an essential strategy in the face of AI-driven changes in the job market. As some jobs become automated or radically transformed by AI, the demand for the skills associated with these jobs may decrease. At the same time, new jobs are emerging, often requiring skills that are different from those needed in traditional jobs.

Consider the example of manufacturing. As automation becomes more prevalent in this industry, some manufacturing jobs are being displaced. However, new jobs are also being created—jobs that involve designing, programming, and maintaining the automated systems. These jobs require skills—such as coding, data analysis, and robotics—that are different from those needed in traditional manufacturing jobs.

Reskilling can be a powerful strategy for helping workers navigate these changes. By learning new skills, workers can transition from declining occupations to growing occupations, minimizing the risk of unemployment and maximizing their chances of career growth.

However, reskilling is not without its challenges. It requires time, effort, and often financial investment. It also requires a growth mindset—the belief that one's abilities can be developed through dedication and hard work.

The Demand for Upskilling

Upskilling, or the process of learning new skills to perform one's existing job more effectively or to take on new responsibilities within the same field, is another important strategy for adapting to the AI-driven changes in the job market.

As AI systems become more integrated into various jobs, there's a growing demand for skills related to AI. For instance, a salesperson might need to learn how to use AI tools for customer relationship management, or a doctor might need to learn how to interpret the outputs of an AI system that analyzes medical images.

Upskilling can be beneficial not only for individual workers, who can enhance their job performance and career prospects, but also for organizations, which can increase their productivity and competitiveness. However, like reskilling, upskilling requires a commitment to lifelong learning and a supportive environment that encourages and enables skill development.

In the next section, we will explore the various approaches to reskilling and upskilling, and discuss how they can be effectively implemented to prepare the workforce for the future of work.

Approaches to Reskilling and Upskilling

As we grapple with the wave of change brought about by AI, it becomes clear that our approach to education and skill development must also evolve. Traditional models of education, where learning is concentrated in the early years of life and primarily takes place in formal educational institutions, may not be sufficient for the rapidly changing job market. We need flexible, accessible, and lifelong learning opportunities that enable individuals to continually update their skills and adapt to new challenges and opportunities.

Reskilling Initiatives

There are various approaches to reskilling, ranging from formal education programs to online learning platforms.

Formal education programs, such as degree programs or vocational training programs, can provide a structured and comprehensive pathway to acquiring new skills. However, they can be time-consuming and expensive, and may not be accessible to everyone, particularly those who are already in the workforce.

Online learning platforms offer a more flexible and accessible alternative. Platforms like Coursera, edX, and Udacity offer a wide range of courses in various fields, often for free or at a low cost. These platforms allow individuals to learn at their own pace, at any time and from anywhere. They also offer credentials, like certificates or microdegrees, that can signal the individual's new skills to employers.

Reskilling bootcamps are another promising approach. These are intensive training programs, typically lasting a few weeks to a few months, that aim to equip individuals with new skills quickly. Bootcamps are particularly popular in fields like coding and data science, but they are expanding into other fields as well.

Upskilling Initiatives

Upskilling can be achieved through similar means as reskilling, such as formal education programs, online learning platforms, and bootcamps. However, upskilling also often takes place on the job, through professional development programs, mentoring, and on-the-job training.

Employers play a crucial role in upskilling. By providing upskilling opportunities, employers can not only enhance their employees' job performance and productivity, but also increase employee engagement and retention.

However, not all employers have the resources or the will to invest in upskilling. This is where public policy can play a role, by providing incentives for employers to invest in upskilling, or by directly providing upskilling opportunities through public programs.

Challenges and Opportunities

Reskilling and upskilling are not without their challenges. They require a significant investment of time and effort, and often financial resources as well. There may also be psychological barriers to overcome, such as fear of failure or resistance to change.

However, with the right support and resources, reskilling and upskilling can open up new opportunities for individuals and societies. They can help individuals to stay relevant in the job market, enhance their career prospects, and achieve personal growth. At a societal level, they can help to mitigate the impacts of automation, foster economic growth, and promote social cohesion.

As we look to the future, it is clear that reskilling and upskilling will be critical strategies for navigating the AI-driven changes in the job market. By embracing lifelong learning and fostering a culture of adaptability, we can prepare ourselves and our societies for the opportunities and challenges that lie ahead.

In the next section, we will explore how AI is not just changing existing jobs, but also creating new ones. We will look at the new roles and industries that are emerging as a result of AI advancements, and discuss what they mean for the future of work.

Part III: AI as a Job Creator: New Roles and Industries

While AI's potential to automate certain jobs can seem daunting, it is crucial to remember that this is only one side of the coin. Just as previous industrial revolutions have done, the advent of AI brings with it a wave of new roles and industries. As we explore this transformative landscape, we'll see that AI is not just a job taker, but can also be a significant job creator.

Emerging Jobs within the AI Industry

The AI industry itself is a hotbed for new job roles. The increasing demand for AI applications across sectors has led to a surge in roles such as AI specialists, data scientists, and machine learning engineers. These roles involve designing, developing, and maintaining AI systems, and they require a combination of skills in areas like programming, statistics, and domain knowledge.

Consider the role of an AI ethicist, a relatively new job that has emerged in response to the growing recognition of ethical issues in AI. AI ethicists help to ensure that AI systems are developed and used in a way that respects ethical principles, such as fairness, transparency, and respect for privacy. They might work on developing ethical guidelines for AI, advising on ethical issues in AI projects, or conducting research on AI ethics.

AI-induced Jobs outside the AI Industry

Beyond the AI industry, the influence of AI is leading to the creation of new roles in a variety of sectors. These roles often involve interacting with, managing, or making decisions based on AI systems.

For instance, in healthcare, roles are emerging for professionals who can bridge the gap between AI technology and clinical practice. These individuals need to understand both the capabilities and limitations of AI, as well as the specific needs and constraints of healthcare.

In business, as AI becomes more integrated into decision-making processes, there is a growing need for roles that involve interpreting and communicating the outputs of AI systems. This might involve explaining the results of an AI analysis to non-technical stakeholders, or integrating AI insights into business strategies.

Potential of New Industries

AI's potential to create new industries is perhaps its most exciting aspect as a job creator. As AI technologies continue to evolve and find new applications, they have the potential to give rise to entirely new industries that we can hardly imagine today.

Take the autonomous vehicle industry as an example. While still in its early stages, this industry has the potential to transform transportation, logistics, and mobility, leading to a host of new jobs in areas like autonomous vehicle design, fleet management, and regulatory compliance.

Challenges and Opportunities

While the emergence of new roles and industries due to AI is exciting, it also brings challenges. These new jobs often require advanced and specialized skills, which means that education and training systems need to keep up. There's also the risk that these new jobs could exacerbate inequality if they are mainly high-skilled jobs that are out of reach for lower-skilled workers.

Nonetheless, the potential of AI as a job creator provides a counterbalance to the narrative of AI as a job destroyer. As we continue to navigate the AI revolution, it is crucial to keep this balance in mind and to look for ways to maximize the job-creating potential of AI, while minimizing its job-destroying impacts.

In conclusion, the interplay between AI and jobs is complex and multifaceted, involving both challenges and opportunities. As we stand at the precipice of this new era, it is crucial that we approach these changes with a sense of balance, understanding, and foresight. This will allow us to harness the potential of AI to enhance work, create new opportunities, and drive inclusive growth.

CHAPTER 10: THE LIMITATIONS AND CHALLENGES OF AI

As we stand at the crossroads of the Fourth Industrial Revolution, it becomes increasingly important to understand not only the potential of artificial intelligence (AI), but also its limitations and the challenges it presents. Despite the significant advancements in AI, there are still many hurdles that need to be overcome to realize its full potential. In this chapter, we will delve into some of the key limitations and challenges of AI, from technical and ethical issues to societal implications.

Part I: Technical Limitations of AI

The lure of AI lies in its extraordinary potential: the promise of machines that can learn from experience, adapt to new situations, and perform tasks that once required human intelligence. Yet, for all its remarkable progress, AI still faces significant technical limitations. As we venture deeper into the realm of AI, it is critical to recognize these limitations and understand their implications.

Lack of Understanding and Explainability

One of the most formidable challenges in AI, particularly in the realm of deep learning, is the lack of understanding and explainability. Often, AI systems operate as "black boxes," wherein their internal workings are hidden, and their decision-making process is challenging to decipher. This opacity can lead to a host of issues, from erroneous decisions due to hidden biases to potential risks in safety-critical applications like autonomous driving or medical diagnosis.

The issue of explainability in AI is not merely a technical one—it has profound ethical and legal implications. For example, if an AI system makes a decision that has serious consequences—such as denying a loan application or recommending a particular medical treatment—there's a need for transparency so the decision can be understood, justified, and potentially contested. The challenge lies in balancing the complexity and performance of AI models with the need for transparency and interpretability.

Dependence on Data

The performance of AI systems is closely tied to the quantity and quality of data they are trained on. AI models, particularly deep learning models, require large amounts of data to learn effectively. The gathering of such extensive datasets can be a challenging and resource-intensive process. Moreover, the use of personal data raises significant privacy concerns and necessitates stringent data governance.

In addition to the quantity of data, the quality of data is equally important. AI systems learn from the data they are trained on, and if this data is biased or flawed, the systems can perpetuate or even amplify these biases. For instance, an AI system trained on a dataset that underrepresents certain demographic groups might perform poorly for individuals from those groups.

Difficulty in Handling Complex and Novel Situations

AI systems, while remarkably capable in their specific domains, often struggle when faced with complex situations that require a nuanced understanding of the world, or when presented with tasks that are outside their training data.

For instance, AI-powered chatbots can provide customer service or answer FAQs but might falter when asked to provide advice on a complex and novel issue. Similarly, while AI has achieved superhuman performance in games like chess and Go, which have well-defined rules and a limited set of possible states, it struggles in real-world situations that are uncertain, ambiguous, and constantly changing.

In the next section, we will delve into the ethical and societal challenges posed by AI, adding another layer to our understanding of the complexities inherent in this transformative technology.

Part II: Ethical and Societal Challenges

While the technical limitations of AI pose significant challenges, it is the ethical and societal implications that often take center stage in public discussions about AI. As AI systems become increasingly integrated into our lives and societies, it is crucial to address these challenges and ensure that AI is used responsibly and ethically.

Bias and Fairness

One of the most significant ethical challenges in AI is the risk of bias. If the data used to train an AI system reflects societal biases, the system can learn and perpetuate these biases. This can lead to unfair outcomes when the AI system is used for decision-making.

Consider, for example, an AI system used for hiring. If the training data includes a disproportionate number of successful candidates from a particular demographic group, the AI system might learn to favor candidates from that group, leading to biased hiring decisions.

Addressing bias in AI is not a straightforward task. It requires a careful examination of the data used to train the AI system, as well as the algorithm itself. It also requires a nuanced understanding of fairness and a commitment to ensuring that AI systems respect the principle of equal treatment.

Privacy and Security

As AI systems often rely on large amounts of data, including personal data, they raise significant privacy concerns. The use of personal data in AI must be carefully managed to respect individuals' privacy rights and comply with data protection regulations.

Moreover, as AI systems become more integrated into critical infrastructures, they become attractive targets for malicious actors. Ensuring the security of AI systems is therefore of paramount importance. This includes not only the security of the AI system itself but also the security of the data it uses and generates.

Impact on Jobs and the Economy

The economic implications of AI are vast and complex. On the one hand, AI has the potential to drive economic growth by improving efficiency and enabling new products and services. On the other hand, there is a concern that AI could displace jobs, particularly those involving routine tasks, leading to unemployment and economic inequality.

The challenge lies in managing the transition to an AI-driven economy in a way that maximizes the benefits of AI, while minimizing its negative impacts. This requires proactive policies and strategies, such as reskilling and upskilling initiatives, to prepare the workforce for the AI revolution.

In the next section, we will explore potential approaches to navigate these challenges, including regulation and oversight, as well as research and innovation.

Part III: Navigating the Challenges

The challenges posed by AI are as vast and complex as the technology itself. However, these challenges are not insurmountable. With careful planning, thoughtful regulation, and innovative research, we can navigate the AI landscape responsibly and ethically.

Regulation and Oversight

As AI continues to permeate society, the role of regulation and oversight becomes increasingly crucial. Regulation can provide a framework to address some of the ethical and societal challenges posed by AI, such as bias, privacy, and impact on jobs.

However, regulating AI is not a straightforward task. It requires a delicate balance between promoting innovation and protecting societal interests. Too much regulation could stifle innovation and economic growth, while too little could leave room for misuse and unintended consequences.

Current state of AI regulation varies widely across the world, with some countries taking a more hands-on approach, while others favoring a more laissez-faire stance. However, there's a growing consensus on the need for some form of AI regulation, particularly in areas such as data protection, transparency, and accountability.

Research and Innovation

While regulation provides a framework for the responsible use of AI, it is through research and innovation that we can address some of its technical limitations and ethical challenges. This includes research on explainable AI, fair and unbiased algorithms, and secure and privacy-preserving AI technologies.

Innovation in AI also involves developing new applications and use-cases that can bring benefits to society. This requires a multidisciplinary approach that brings together expertise from different fields, such as computer science, social sciences, and humanities.

<u>Conclusion</u>

As we navigate the AI revolution, it is crucial to keep in mind both the potential and the pitfalls of this transformative technology. By understanding its limitations and challenges, and by implementing thoughtful regulation and innovative research, we can ensure that AI is used in a way that benefits society as a whole.

The journey into the AI era is not a solitary one. It requires the collective efforts of all stakeholders—policymakers, technologists, businesses, educators, and citizens—to navigate the challenges and seize the opportunities. As we continue on this journey, let us do so with a sense of purpose, responsibility, and optimism for the future that AI can help us build.

CHAPTER 11: THE INTERSECTION OF AI AND OTHER EMERGING TECHNOLOGIES

In the constantly evolving technological landscape, artificial intelligence (AI) has secured a place of prominence. However, to view AI in isolation would be a narrow perspective. The real magic unfolds when AI intersects with other burgeoning technologies, creating a synergy that propels us further into the realm of possibilities. In this chapter, we will explore the confluence of AI with the Internet of Things (IoT), blockchain, and quantum computing.

Part I: AI and the Internet of Things (IoT)

The Internet of Things (IoT) is a network of interconnected devices that communicate and exchange data with each other. From smart homes equipped with automated temperature control to wearable fitness devices tracking health metrics, IoT has permeated various aspects of our lives. But how does this connect with AI?

Understanding the IoT

The IoT is more than just a collection of devices—it's a vast, interconnected system that generates a treasure trove of data. These devices, embedded with sensors, software, and other technologies, collect and share data about the environment in which they operate. This constant stream of data provides the perfect playground for AI to work its magic.

Synergy between AI and IoT

AI and IoT are a match made in technology heaven. AI, with its ability to analyze large volumes of data and glean insights, can greatly enhance the utility of IoT devices. AI can process the data generated by IoT devices to make predictions, automate decisions, and even adapt device behavior based on the insights gleaned.

For instance, imagine a network of IoT devices deployed across a city to monitor traffic conditions. The sheer volume of data generated by these devices would be overwhelming for humans to analyze. However, an AI system could process this data in real-time, predict traffic congestion, and provide alternative route suggestions to drivers.

Conversely, the IoT can also enhance AI. The data collected by IoT devices can be used to train and improve AI models, making them more accurate and effective. Furthermore, the IoT can provide a platform for AI to operate and make an impact in the real world.

Case Studies

To illustrate the synergy between AI and IoT, let's look at a few real-world examples.

In agriculture, AI and IoT are being used together to optimize irrigation. Sensors deployed in fields measure soil moisture levels and relay this data to an AI system. The AI system analyzes the data, predicts the optimal irrigation schedule, and automatically controls the irrigation equipment accordingly. This combination of AI and IoT can save water, improve crop yields, and reduce labor costs.

In healthcare, AI and IoT are revolutionizing patient monitoring. Wearable devices can continuously monitor a patient's vital signs and other health metrics, and an AI system can analyze this data to detect anomalies. If the AI system detects a potential health issue, it can alert the patient and their healthcare provider, enabling early intervention.

As we move forward, the intersection of AI and IoT holds immense potential to transform various sectors and improve our lives in ways we can only begin to imagine.

In the next section, we will explore another exciting intersection —that of AI and blockchain.

Part II: AI and Blockchain

Blockchain, most known as the technology underpinning cryptocurrencies like Bitcoin, is essentially a decentralized and distributed digital ledger of transactions. This technology offers transparency, security, and immutability, making it attractive for applications beyond finance, such as supply chain management and healthcare. When we combine the analytical prowess of AI with the secure architecture of blockchain, we create a potent mix that can address some of the current challenges faced by both technologies.

Understanding Blockchain

A blockchain is a chain of blocks, with each block recording a batch of transactions. The beauty of blockchain lies in its security design—once a block is added to the chain, it is nearly impossible to change the information within it. This is due to its decentralized nature, where the same blockchain is maintained on multiple computers in a network, and a majority must agree for any changes to be made. This makes blockchain resilient to tampering, providing a high level of trust and security.

Synergy between AI and Blockchain

The intersection of AI and blockchain opens up new possibilities for both technologies. Blockchain can provide a secure and transparent infrastructure for training AI models. Using blockchain, the data used to train AI models can be traced back to its source, providing transparency and ensuring data integrity. This can be particularly useful in sectors like healthcare, where the provenance of data is crucial.

On the other hand, AI can enhance blockchain systems. For instance, AI algorithms can optimize the performance of blockchain networks, making them more efficient and scalable. AI can also be used to automate the process of verifying and adding transactions to the blockchain, which is currently a resource-intensive process.

Case Studies

The combination of AI and blockchain is being explored in various sectors. One promising application is in data marketplaces. Blockchain can provide a secure and transparent platform for individuals to sell their data, and AI algorithms can analyze this data to generate insights. This can democratize access to data and provide individuals with control and compensation for their data.

Another application is in supply chain management. IoT devices can capture data about goods as they move through the supply chain, and this data can be recorded on a blockchain, providing a transparent and tamper-proof record. AI can then analyze this data to optimize the supply chain, predict and manage risks, and ensure the authenticity of goods.

In the next section, we will dive into the intersection of AI and another groundbreaking technology—quantum computing. This powerful combination has the potential to take AI to new heights, unlocking capabilities we've only begun to dream of.

Part III: AI and Quantum Computing

As we explore the intersections of AI with other technologies, we cannot overlook the burgeoning field of quantum computing. Quantum computing, still in its early stages, has the potential to revolutionize computing by performing complex calculations at speeds far beyond the capabilities of classical computers. This could dramatically accelerate the development of AI and open up new possibilities.

Understanding Quantum Computing

Quantum computing leverages the principles of quantum mechanics to process information. While classical computers use bits as their smallest unit of data, with each bit being either a 0 or a 1, quantum computers use quantum bits, or qubits. What sets qubits apart is their ability to be in a state of superposition, meaning they can be both 0 and 1 at the same time. This allows quantum computers to perform multiple calculations simultaneously, providing a computational speed that classical computers cannot match.

Synergy between AI and Quantum Computing

The integration of AI and quantum computing could lead to a quantum leap in AI's capabilities. Quantum computing could significantly speed up AI computations and enable the processing of larger datasets, leading to more accurate and sophisticated AI models.

For instance, the training of deep learning models, which requires a significant amount of computational power, could be greatly accelerated by quantum computing. This could make it feasible to train more complex models, or to use larger and more diverse datasets for training, thereby improving the performance of the AI system.

Conversely, AI can also enhance quantum computing. For example, AI algorithms can be used to optimize the performance of quantum computers, such as by managing errors, which are a significant challenge in quantum computing.

Case Studies

While the intersection of AI and quantum computing is still a nascent field, there are some promising early applications.

For instance, Volkswagen has used a quantum computer to develop a traffic flow optimization system. The system uses data from thousands of taxis to calculate the fastest route for each taxi, taking into account real-time traffic conditions. The complexity of this problem makes it difficult to solve with classical computers, but a quantum computer can perform the necessary calculations efficiently.

Another application is in drug discovery. Quantum computers could potentially simulate the molecular structure of new drugs and predict their effects, a task that is currently computationally infeasible for classical computers. Combined with AI, this could significantly speed up the drug discovery process and lead to the development of new treatments.

Conclusion

The intersections of AI with other emerging technologies such as IoT, blockchain, and quantum computing hold immense potential. They represent the next frontier of innovation, where the whole is greater than the sum of its parts. As we continue to explore these synergies, we must do so with a sense of responsibility and foresight, recognizing the potential implications and challenges. It is through such a balanced and thoughtful approach that we can harness these powerful technologies to create a future that is not only technologically advanced, but also socially and ethically responsible.

CHAPTER 12: THE JOURNEY THROUGH AI AND BEYOND

As we turn the final pages of this exploration, we stand at a vantage point, looking back at the terrain we've traversed and gazing into the horizon of possibilities that AI, in conjunction with other technologies, presents.

Our journey began with a deep dive into the world of AI tools, where we discovered how algorithms learn from data, how neural networks mimic the human brain, and how reinforcement learning allows machines to learn from interaction with their environment. We delved into the realm of natural language processing, computer vision, and robotics, exploring the breadth and depth of AI's capabilities. We demystified AI's complex concepts and provided hands-on exposure to AI tools, setting a strong foundation for understanding this transformative technology.

But understanding the tools was only the beginning. We went further, exploring the ethical landscape of AI, where we grappled with questions of fairness, transparency, and responsibility. We recognized the need for a human-centric approach to AI, one that prioritizes the respect for human rights and the promotion of societal well-being.

We then navigated the intricate relationship between AI and jobs, witnessing the dual faces of AI as both a disruptor and creator of jobs. We understood the necessity of reskilling and upskilling, the value of human-AI collaboration, and the potential of AI to drive economic growth and societal progress.

Next, we confronted the limitations and challenges of AI, acknowledging that despite its immense potential, AI is not a panacea. We discussed the technical hurdles, the ethical dilemmas, and the societal implications, realizing that our journey with AI will be one of continuous learning, adaptation, and careful navigation.

In our final leg, we ventured into the intersections of AI with other emerging technologies—IoT, blockchain, and quantum computing. We glimpsed the synergies that these combinations could create, opening up new avenues for innovation and addressing some of the current challenges faced by AI.

As we conclude, we are left with a sense of awe for the vast potential of AI and a sober recognition of the challenges that lie ahead. We understand that the journey with AI is not a sprint but a marathon, requiring persistence, continuous learning, and a keen eye for the ethical and societal implications of the technology.

The story of AI is still being written, and we all have a role to play in shaping its narrative. As we step into the future, let us do so with a commitment to harnessing AI for the greater good, to cultivating the skills needed for an AI-driven world, and to creating a future where AI and humanity coexist and thrive.

Thank you for joining me on this journey through the fascinating world of AI. As we part ways, I leave you with a quote by Eliezer Yudkowsky: "By far, the greatest danger of Artificial Intelligence is that people conclude too early that they understand it." So, let's keep exploring, keep questioning, and keep learning. After all, the journey is just beginning.

ACKNOWLEDGEMENTS

Authored by Vihang Chheda
Technical support by Raj Global
Cover Art by Ishita Agrawal